Cutting-Edge Advances in Hydraulic Engineering

Edited by Harry Edmar Schulz and André Luiz Simões

Published in London, United Kingdom

Cutting-Edge Advances in Hydraulic Engineering
http://dx.doi.org/10.5772/intechopen.1006432
Edited by Harry Edmar Schulz and André Luiz Simões

Contributors
Aida Aslanova, André Luiz Simões, Carlos Eugenio Pereira, Francisco Osny Enéas da Silva, Fuad Veliev, Harry Edmar Schulz

First published in London, United Kingdom, 2025 by IntechOpen
IntechOpen is the global imprint of INTECHOPEN LIMITED, registered in England and Wales, registration number: 11086078, 167-169 Great Portland Street, London, W1W 5PF, United Kingdom

For EU product safety concerns: IN TECH d.o.o., Prolaz Marije Krucifikse Kozulić 3, 51000 Rijeka, Croatia, info@intechopen.com or visit our website at intechopen.com.

British Library Cataloguing-in-Publication Data
A catalogue record for this book is available from the British Library

Cutting-Edge Advances in Hydraulic Engineering
Edited by Harry Edmar Schulz and André Luiz Simões
p. cm.
Print ISBN 978-1-83634-935-8
Online ISBN 978-1-83634-934-1
eBook (PDF) ISBN 978-1-83634-936-5

If disposing of this product, please recycle the paper responsibly.

Meet the editors

Harry Schulz holds a Bachelor's degree in Civil Engineering and Philosophy, as well as an MSc and a Ph.D. in Hydraulics and Sanitation. He also graduated in Mechanical Engineering with a one-year Certification in Transport Phenomena Education. He was awarded the title of Full Professor in Transport Phenomena and Fluid Mechanics at the University of São Paulo. He spent two years as a Visiting Professor at the Institut für Hydromechanik, Universität Karlsruhe (Germany), and one year as a Visiting Professor at both the Saint Falls Laboratory, University of Minnesota (USA), and the Department of Civil and Environmental Engineering, University of Alberta (Canada). He is currently a Lecturer at Auburn University, USA. His former positions at the University of São Paulo include Director of the Centre for Water Resources and Applied Ecology, Head of the Department of Hydraulics and Sanitation, and Coordinator/Vice-Coordinator of the Graduate Program in Hydraulics and Sanitation.

André Simões holds a Ph.D. and MSc in Hydraulic and Sanitation Engineering from the University of São Paulo (USP) and a BSc in Civil Engineering from the University of Salvador. Since 2014, he has been an Adjunct Professor at the Federal University of Bahia (UFBA), where he teaches Hydraulics and Advanced Computational Fluid Dynamics (CFD) at both undergraduate and graduate levels. He served as Coordinator of UFBA's Graduate Program in Environment, Water, and Sanitation (MAASA) from 2019 to 2022 and has been Vice-Coordinator from 2023 to March 2025. His research focuses on fluid mechanics, multiphase flows, CFD, and hydraulic structures. He leads the Computational Fluid Dynamics and Hydraulics Group at UFBA and is also a member of the Solid Waste and Effluent Research Group and the Optical Properties Laboratory at UFBA's Institute of Physics.

Contents

Preface

The primary purpose of this book is to present the reader with an engaging collection of real-world problems in hydraulics, along with a thorough exposition of the procedures and theories employed for their analysis, solution, and application in engineering. To achieve this objective, this work counted on the direct effort of six researchers who provided five chapters covering their expertise, and which embrace the working space-scales of: i) nano scale; ii) porous scale; iii) human scale (city); and iv) natural scale (hydrologic basis, rivers). The presented results were also shaped by the contributions of colleagues and students working with these researchers, whose help is valued and appreciated.

The authors detailed their experimental methodologies and theoretical considerations in a way that facilitates the reader's transit through the main aspects of each study. Equations, photographs, explanatory diagrams, and graphs are adequately presented alongside the discussions of the methodologies and the obtained results.

The studies were applied to specific situations, but the application of the results and methods can be expanded to new situations, considering the basic assumptions that were applied.

As the Academic Editors, we are confident that this book will contribute to the search for solutions to new problems in the evolving field of hydraulics.

Harry Edmar Schulz
Department of Civil and Environmental Engineering,
Samuel Ginn College of Engineering,
Auburn University,
Auburn, USA

André Luiz Simões
Department of Environmental Engineering,
Polytechnic School,
Federal University of Bahia,
Salvador, Brazil

Section 1

Introduction

Chapter 1

Introductory Chapter: Hydraulics and Its Various Scales of Application

Harry Edmar Schulz and André Luiz Simões

1. The different points of view in hydraulics

Considering our external environment together with the internal characteristics of our bodies, we depend on fluids externally and internally, on their movements, and on their capacity to transport components valuable to our day-to-day activities.

Illustrating the former paragraph, as living beings on Earth, our natural condition imposes our total immersion in a fluid medium: the air. Additionally, our bodies conduct nutrients and oxygen internally through a proper liquid (the blood), which is hydraulically pumped during our whole lives (the heart).

It is thus natural to study in more detail the fluids and their movements, intending to use them in increasingly optimized ways. In this sense, for example, the present industrial production needs involve the use of small-scale technological advances, which imply studying fluids in geometries of nano-environments, that is, attaining very small space scales for which only recently useful information is being generated [1].

In the other extreme, the natural phenomena observed in our globe involving fluids include, for example, intense storms and rains that result in large floods, erosion patterns, and relocations of soil or sediment. The interactions of rainwater and soil imply in continuous modifications of the known hydrologic basins, which generate the need for continuous measurements, proposing and applying new equipment and methodologies to monitor our living spaces [2].

The interaction of rain events and the hydrological basins involves floods, evaporation, transpiration, and absorption of the water by the soil. Each of these phenomena has its own characteristics and difficulties for precise quantification. The quantification of the absorption and flow of water through the soil is still being developed by different researchers around the globe, generating knowledge that is also applied to artificial soil or gravel structures. One useful tool to control flows in hydrological basins is the gabion cage, the design of which directly benefits from the mentioned knowledge of flow through porous media (or soil) [3].

Considering the human scale, that is, our immediate outer environment, our cities are built in sites of the hydrologic basins that were considered the most adequate for the settlement of the previous generations. The evolution of our societies in terms of number and quality of life has also brought the need to maintain our immediate outer environment adequately clean. However, the collateral effect is that we did not "destroyed" the "dirt" or "filth" of our cities, but we "exported it" to "out of the cities" generating an uncontrolled pollution of the more distant surroundings.

IntechOpen

Focusing on the built environment of our cities, and not on their surroundings, the modifications imposed on the coverage of the soil of the basin while building and expanding the urban areas, together with the transformation of natural creeks into tube flows, generated new kinds of problems for these built environments. For example, high pressure flows that generate very high velocities in drainage systems, together with artificial geysers during rain events, are typical phenomena of built environments, which can even destroy the built structures that formed them [4].

Finally, either caused by long-term solar cycles or by human irresponsible activities, the already registered global heating points to modifications in the time and space distribution of the water cycle and, consequently, in its use as historically performed in the different parts of the globe and in the different human activities. As a consequence, we need studies that indicate new and optimized solutions to better use our water reserves and the other fluids that help our survival in adequate living conditions.

2. Continuous studies on different scales

Scientists, researchers, and engineers working in theoretical, experimental, and numerical studies know the urgency of a better understanding of the changing conditions and the need for adequate ways to solve present and future problems. New procedures, or even the use of existing procedures in new situations and conditions, must be exposed and tested [5].

Being aware of this urgency, the present book on "Hydraulic Engineering – Cutting Edge" brings together studies of professionals dealing with water flows in its

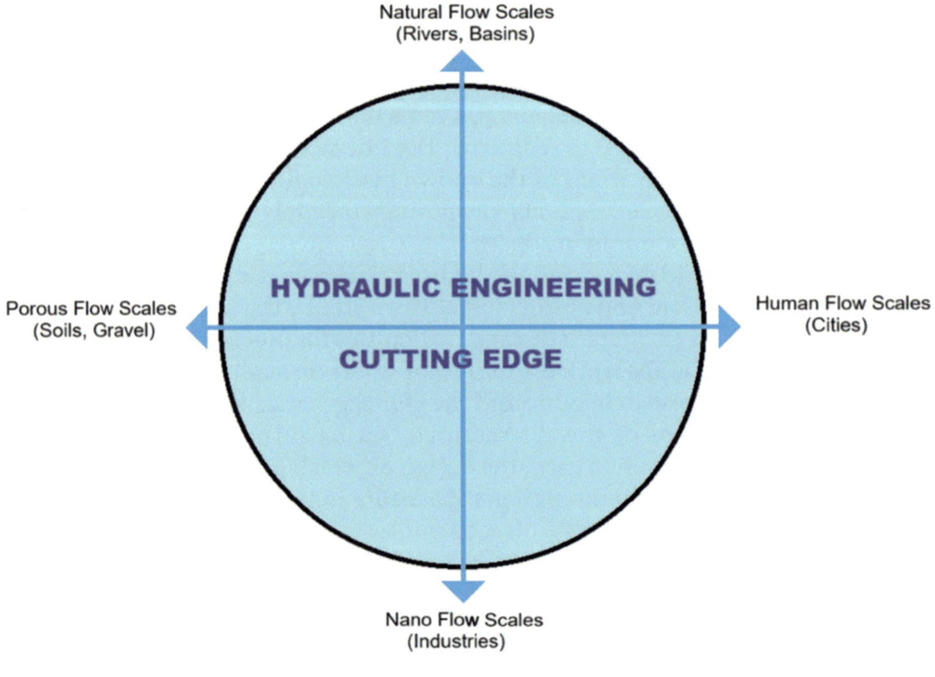

Figure 1.
The spectrum of scales covered by this book: from the smaller dimension to the largest dimension, implying an evolution from low to high velocities.

DOI: http://dx.doi.org/10.5772/intechopen.1012394

different scales: nano-scale, porous scale, human scale, and natural scale. **Figure 1** shows the four aspects covered by this book. Quantitative aspects of questions related to the mentioned scales in Hydraulic Engineering are adequately described in the following chapters.

Author details

Harry Edmar Schulz[1*] and André Luiz Simões[1,2]

1 Auburn University, Alabama, United States of America

2 Federal University of Bahia, State of Bahia, Brazil

*Address all correspondence to: hes0071@auburn.edu

References

[1] Veliyev FH, Aslanova AR. Non-Newtonian characteristics of water flow in microchannels. Indian Journal of Science and Technology. 2023;**16**:4605-4611. DOI: 10.17485/IJST/v16i47.2691

[2] Pereira CE, Peixoto RAO, Salla MR, Alamy Filho JA, Cunha AP. Application of measurement methods for solid discharge in the Ribeirão Jordão stream and tributaries in UPGRH PN-1. International Journal of Hydrology. 2021;**5**(4):170-180. DOI: 10.15406/ijh.2021.05.00278

[3] Simões ALA, França TIS, Queiroz LM, Lima Neto IE, Schulz HE, Fontes DCH, et al. Dimensionless equations for hydraulic jump stilling basins downstream of gabion stepped chutes. Journal of Hydraulic Structures. 2024;**11**(1):51-64. DOI: 10.22055/jhs.2024.47500.1313

[4] Silva FOE, Palácio FFR Jr, Campos JNB. Rainfall equation for the city of Fortaleza based on pluviograph data from UFC (text in Portuguese). Journal of the Department of Waters and Sewers (Revista DAE). 2013;**192**(May–August):48-59. DOI: 10.4322/dae.2014.106

[5] Schulz HE, Simões ALA, Lobosco RJ, editors. Hydrodynamics Advanced Topics. London: InTech; 2011. DOI: 10.5772/2372

Section 2

Nano Flow Scales and Porous Flows Scales

Chapter 2

Electrokinetic Regulation of Fluid Flow in Microchannels

Fuad Veliev and Aida Aslanova

Abstract

The development of low-permeable hydrocarbon reservoirs is becoming an increasingly urgent task, and therefore, the study of laws of fluid movement in microcracks is a crucial scientific and technical problem. In this book chapter, rheophysical aspects of non-Newtonian behavior of hydraulic systems during flow in thin rectangular channel are considered experimentally. Using the microchannel model, it is established that the nonlinear rheological effect in the fluid flow in micro-slits is mainly caused by the value of the electrokinetic potential of the system, by reducing of which it is possible to significantly weaken the non-Newtonian nature of the fluid. To regulate electrokinetic potential of fluid system, antistatic additives were used; the optimal concentration of which was defined experimentally. Based on the Bingham model, the rheological parameters of fluid flow were estimated at different micro-slit clearances, in the absence and presence of an antistatic additive. It is also established that a reduction in electrical potential of fluid flow leads to a significant decrease in yield shear stress during the fluid flow in the microchannel. The role of electrokinetic potential in Pressure Recovery Curves (PRC) of hydraulic system and oil displacement process by water in the microchannels has also been studied experimentally.

Keywords: antistatic additives, slit openness, electric double layer, streaming potential, microchannel, yield shear stress

1. Introduction

Over the past 20 years, the development and exploration of unconventional low and ultralow permeable oil and gas resources have been one of the main demands of the oil industry and global energy supply throughout the world [1].

Currently, 10% of the world's proven reserves account for low and ultralow permeable hydrocarbon deposits. Low and ultralow permeable oil and gas fields contain the main hydrocarbon resources of many countries over the world as the USA, Canada, China, Argentina, Russia, Mexico, Arabia, and Azerbaijan [2]. On average, the recovery factor of these rocks is 20%, and most of the reserves remain underground as residual reserves [3].

One of the main difficulties arising in the development and exploration of those reservoirs is the manifestation of anomalies in the rheophysical properties of fluid flow in low-permeable and microcrack formations. In low and ultralow permeable formations, the linear dependence of the filtration process is violated during the movement of viscous formation fluids in the contact zone of the liquid and rock, which, accordingly, leads to a violation of Darcy's law [4, 5]. At the same time, it was found that as the openness of microcracks decreases, starting from a certain critical thickness, non-Newtonian properties are observed with the manifestation of the initial pressure gradient in a viscous liquid (water, oil). Since these formations are very sensitive to pressure changes, the manifestation and increase in the initial pressure gradient leads to a sharp decrease in the permeability of the deposits, which results in a significant weakening and even blocking of the flow [6, 7].

The rheophysical properties of fluid flow in microchannels differ significantly from the properties of fluid flow in macro-dimensional channels and porous media. This difference is fundamentally due to the fact that since the size of the liquid-rock interface in microchannels is so large that it can be compared with the total flow volume, intermolecular forces dominate the regulation of fluid flow [8].

As the size of the microchannel decreases, the value of the flow parameters, as well as the pressure gradient, increases more sharply. This results in the manifestation of non-Newtonian properties of Newtonian fluids in the microchannel [9].

This anomaly can be explained by electrokinetic effects that manifest themselves in the contact zone of the liquid with the wall in microchannels. At the liquid-wall interface, redistribution of charged particles occurs under the influence of intermolecular forces. Thus, charged particles in the liquid solution are attracted by the solid surface and create a motionless liquid layer on the surface of the wall. In this case, redistribution of charged particles occurs under the influence of an electric field applied to the system. Depending on the mechanism of redistribution, various electrokinetic effects manifest at the liquid-wall interface [10].

Detailed study of fluid flow in small and ultrasmall pore environments is of great importance in selecting optimal development technologies for unconventional reservoirs such as shale gas and tight oil. Diwu et al. [11], as a result of their studies in such reservoirs, found that formation water and hydrocarbons have strong non-Newtonian properties and cause the fluid to be blocked and stagnant in the porous medium. Chen et al. [12] found that the manifestation of non-Newtonian properties of the flow significantly affects the well operation process, as well as the efficiency of the water injection process. The issue of optimizing the operation process of these reservoirs requires a comprehensive study of non-Newtonian fluid dynamics in small and ultrasmall pore environments and the development of accurate models. Since such rocks mainly consist of micro- and nano-sized pores, the abovementioned laboratory methods are not sufficient to obtain accurate results [13–15].

On the other hand, the rapid development of microelectromechanical technologies makes it possible to develop new approaches such as studying heat transfer and fluid flow in microchannels. Wu and co-authors [15] applied microelectromechanical technologies to study the dynamics of fluid flow over ultrasmall layered rock samples and discovered new features in the flow. They initially found that in ultrasmall porous media, at low pressure gradients, polarized water molecules are adsorbed on the hydrophilic surface of solid rock. At the same time, the co-authors discovered the exceptional role of boundary layer theory in regulating the nonlinear filtration process of fluid in rock samples. Thus, as the pressure gradient increases, the thickness of the boundary layer decreases up to a certain value, and after that thickness, it remains constant despite the increase in the pressure gradient.

Since 2000, Majid Bahrami, a professor at Simon Fraser University in Canada, and his co-authors have conducted extensive and systematic research on the detailed study of fluid flow in microchannels [6, 16–22].

Bahrami and co-authors studied fully formed laminar fluid flow in mini- and microchannels with smooth surfaces. In many of his works, Bahrami studied the calculation of pressure drop in mini- and microchannels with different cross-sections and its effect on flow characteristics, and developed special models.

Bahrami and his co-authors found that during fluid flow in microchannels, the Poiselli number $f\,\mathrm{Re}_{\sqrt{A}}$ depends on the polar moment of inertia, the area of the cross-section of the thread, and its perimeter [6]. For the ellipse-shaped model, the hydraulic resistance coefficient and Reynolds number were modeled. Both parameters depend only on the geometric dimensions of the model and are able to be applied to different types of flows, regardless of the flow parameters.

Bahrami and his co-authors confirmed the validity of the proposed models in practice by comparing them with the results of research work carried out by Lui and Garimella, Wu and Cheng, and Ga and their co-authors on microchannel models of various shapes. The relative error between the results does not exceed 8%.

In parallel, as a result of his research work on microchannels whose surface is not smooth, Bahrami and his co-authors [19] studied the effect of the degree of smoothness of the surface of the microchannels on the flow parameters and developed an analytical model to detect and predict arbitrary irregularities present on the surfaces. The experiments were carried out on microchannel models with a circular cross-section. Experimental fluid is a noncompressible Newtonian fluid with stable rheological parameters. The flow is in a fully formed laminar mode. It has been established that the microchannel does not have a significant impact on the pressure drop if the surface roughness rate is below 3% and may not be taken into account. And at the value above 3%, it leads to a sharp increase in pressure drops. The proposed model accurately determines the effect of microchannels on flow at high values of surface irregularities and can calculate the pressure drop. However, at low values of surface irregularities, the increase in pressure drop is around 10%, and it is impossible to accurately predict the increase in pressure drop due to the coincidence with the operating accuracy of this model. The model was tested with experimental results on microchannels with high surface irregularities and the processing accuracy was confirmed.

Although there are various approaches to explain the anomalous hydrodynamic properties manifested during the movement of viscous liquids in low-permeable porous media and microcracks, so far, there is no consensus on the mechanism of these phenomena.

The current situation requires the development of new approaches to the study of the filtration processes of fluids and the patterns of their movement in low-permeable microcrack formations.

In none of the existing scientific works in this area, the effect of the electrokinetic potential of the system on the rheophysical properties of fluids, as well as on the flow parameters in porous media and microcracks with low and ultralow conductivity, was taken into account.

The main object of the research work was, on the basis of the microcrack model, the development of a new approaches for studying and eliminating anomalous properties manifested in the rheophysical properties of viscous liquids, considering the influence of electrokinetic factors.

Based on the presented experimental studies, anomalous properties of Newtonian fluids (water) manifested during the flow in the microchannel model (transition of Newtonian liquid to non-Newtonian liquid) were studied by stationary and nonstationary methods and new approaches were offered for significantly weakening anomalous properties of fluids, intensifying the flow properties and increasing the effectiveness of the displacement process by reducing the electrokinetic potential of the system.

2. Electrokinetic processes in hydraulic systems

Real oil and gas condensate systems are heterogeneous, multicomponent, and also consist of water, asphaltenes, resins, paraffin, and other components. According to the well-known Cohen rule of electrostatics, "when two substances with different dielectric conductivity are in contact, an electrical charge occurs in them-a substance with a higher dielectric conductivity is positively charged, and the other is negatively charged" [23]. According to this rule, an electrical double layer (EDL) is formed at the boundary of the components in contact, and thus, a real heterogeneous hydrocarbon system is manifested in the form of a complex composition of an infinite number of local ion-electrostatic microfields distributed in the volume of the dispersion medium. In this composition, the EDL is formed in oil-water, oil-gas, condensate-water, water-gas, paraffin-water, and other boundary surfaces [24].

When a heterogeneous multicomponent hydraulic system is affected by physical fields, a dipole moment is created in some components, while initially, chaotically oriented polar components are directed along certain lines of force, polarization occurs in the EDL, and the system as a whole is no longer electro neutral [25].

EDL can be formed on the basis of three different mechanisms: (1) selective adsorption of an ion of an electrolyte located in solution on a dispersed phase particle. In this case, ions are adsorbed, which, entering the crystal lattice of the particle, cause it to grow and either form a difficult-to-dissolve compound with ions of the opposite sign, or, replacing the corresponding ion in the crystal lattice of the particle, form an isomorphic compound. (2) ionization of molecules on the surface of Dispersed Phase particles. In this case, an EDL is formed due to the dissociation of its molecules on solid phase particles into ions in a liquid dispersed medium. An ion remains on the surface of the particle, which is more firmly bound to it, and an ion that is weaker bound passes into a dispersed medium. (3) while the particles of the system are not capable of exchanging charges, the EDL is formed as a result of the orientation of polar molecules located at the interface. In this case, of the two phases in contact, the one with greater dielectric penetration is positively charged [26]. The distribution of ions in a double electrical layer is complex in nature, and there are various theoretical models of the structure of this distribution.

Initially, it was believed that the EDL is plane and, it was likened to a kind of charged condenser, one of the plates of which consists of the surface of a solid body, and the other is a layer of liquid in contact with this surface (Helmholts model). In this model, the distance between the layers was adopted in the formulation of the molecule diameter, and only the mutual electrostatic forces between the ions were taken into account in the distribution of charges. Under the influence of these forces, the opposite ions try to settle close to the ions adsorbed on the surface of the particles. As a result, two ionic layers are formed, one of which is located on the surface of the particles, and the second in solution. According to the Helmholtz model, the potential changes based on the rectilinear function [27].

Another theory that more precisely expresses the structure of the EDL is the Gui-Chapman theory. According to this theory, the structure of the EDL is determined by the electrostatic attraction between ions and the thermal motion of the particles. Counterions, in addition to accumulating on the surface and forming a monoionic layer, also spread over a certain distance in the liquid phase. Such a distribution of ions is characterized by two opposite factors.

Thus, under the influence of electrostatic attraction forces, ions with opposite signs are attracted to the surface, and as a result of thermal movement (diffusion), they try to spread throughout the entire volume of the liquid phase. As they move away from the interphase boundary, the effect of the electric field gradually weakens and, as a result of diffusion, the density of counterions approaches the density inside the liquid phase. Thus, an equilibrium diffusion layer of counterions in contact with the solid phase is formed. Such an equilibrium state of the diffusion layer is dynamic in nature. On the other hand, ions with the same sign as the potential-determining ions adsorbed on the solid surface are pushed out of the solid phase by electric forces and pass into the liquid. These results determine the distribution of potential-determining ions and counterions in the diffusion layer of the EDL [28].

The Qui-Chapman theory treats ions as point charges. This creates difficulties in the accurate study of the effect of various ions on the EDL and potential, as well as in the explanation of the recharge phenomenon [29].

The currently accepted model, in which these contradictions are also resolved, is the Stern model. This theory combines the Helmholtz-Perrin and Qui-Chapman theories. Stern proceeded from two assumptions when putting forward this theory. First, he assumed that ions are not point charges, they have a certain size, and therefore, their centers cannot be closer to the surface of the solid phase than the distance of the ion radius. Secondly, Stern took into account that in addition to the electrostatic interaction of ions with the surface of the solid phase, there is also a specific adsorption effect [29].

According to Stern's theory, there are two layers of counterions; (1) Adsorption layer: This layer is in contact with the solid surface due to adsorption and electrostatic forces. As a result, a part of the counterions is held by the surface at a very close distance, approximately 1–2 molecule sizes, and forms a plane capacitor with a thickness of σ. The electric potential inside the adsorption layer decreases sharply, as in a plane

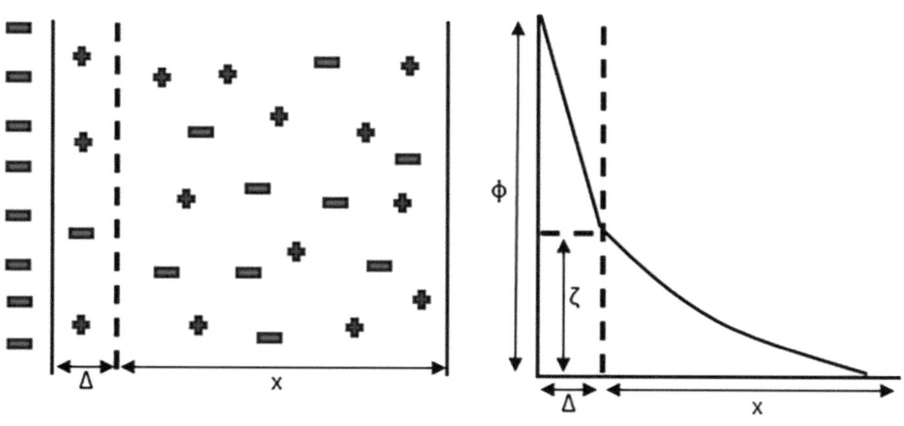

Figure 1.
Electric double layer model.

capacitor, according to a linear function. (2) Diffusion layer: This layer, formed by the counterions spread as a result of diffusion, is in contact with the solid surface only due to electrostatic forces. The potential in this part of the EDL decreases gradually and according to a nonlinear function [29–31].

Thus, according to the modern approach, the IET consists of two main parts. One of them is a stationary layer of ions (Helmholtz layer) that is in direct contact with the surface of the solid, and the other is a diffusion layer that moves when the liquid moves relative to the solid surface (**Figure 1**).

3. Methods and materials

For the experimental study of the flow of Newtonian and non-Newtonian fluids in microchannels, a microchannel model consisting of two steel plates was used. An ultrasmall-sized slit with a length of 30 cm and a width of 40 mm is placed on the inner surface of the steel plates. In addition, the experimental installation consists of a high-pressure gas cylinder, a regulating conductor, a container for storing liquid, a microchannel model with an ultrasmall slit, a pressure gauge for measuring the difference in pressures at the inlet and outlet of the microchannel model, a thermostat for ensuring constant temperature conditions, and thermometers (**Figure 2**).

Lead papers were used to adjust the openness of the crack of the microchannel model. First of all, the plates of the model were placed on top of each other and connected by bolts and the thickness of two plates together from all four sides of the model was measured. Then, bolts were unscrewed, and lead papers of different thicknesses were alternately placed between the plates, which were pre-measured, and bolts were reattached. The thickness of the plates was re-measured together with lead paper. The difference between two thicknesses indicates the openness of the slit. The defined thickness difference gives the same result with the thickness of the lead paper. The accuracy of the openness of the slit with this method is in the range of 5×10^{-7} m.

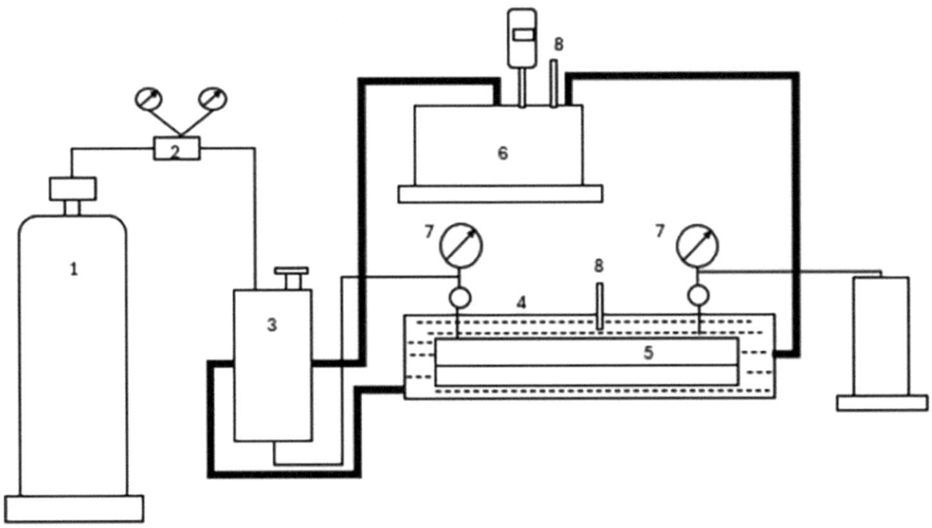

Figure 2.
Scheme of the experimental installation. 1-high-pressure gas cylinder, 2-regulating conductor, 3-container, 4-Thermo bath, 5-microchannel model, 6-thermostat, 7-monometer, 8-thermometer.

Through the experiments, the openness of the plates was repeatedly checked and no deformation was observed [5].

In order for the liquid to enter the slit of the model and be able to leave the slit, an additional slit (40 × 5 × 1) mm in size was placed in the inlet and outlet sections of the bottom plate.

The air permeability of the microchannel model was determined by the classical method. It has been established that the ratio of $\sqrt{k}/h$ is equal to 0.56.

As it is known in the method of determining the permeability in a porous medium by air, the obtained permeability value can be significantly greater than the real permeability due to the manifestation of the Klinkenberg effect. In order to verify the accuracy of the permeability value calculated by this method, the permeability for the liquid was determined with the extrapolation at the openness of $h = 5$ μm. It has been established that the permeability value measured by air is the same as the permeability for a liquid calculated by the extrapolation method. For the exact determination of permeability, similar experiments were also carried out at the values of 15, 20, and 25 μm of openness, and the uniformity of the permeability values obtained by two different methods was confirmed.

One of the main factors affecting the accuracy of the results is the complete saturation of the model with liquid. To ensure the high accuracy of this process, the model is vacuumed before it is saturated with liquid and the empty weight of the model is measured with an accuracy of ±0.0001 G. The liquid is pumped into the crevice under a low pressure gradient and the model is weighed again along with the liquid. This process is repeated many times until the weight of the model together with the liquid reaches a fixed value. At all values of the pressure drop between the inputs and outputs of the microchannel model, the flow rate of the liquid is repeatedly recorded. If the flow rate no longer changes depending on the time, it is assumed that the flow has reached a stable mode. After the flow stabilizes, the flow rate is set at different times, not less than four times. The weight of the glass container empty and filled with liquid, which is used to measure the mass of the liquid that leaves the slit, is measured. The mass difference in the empty and full state of the glass container indicates the mass velocity of the liquid flow. As the final flow rate, the numerical mean of the recorded results is taken. This process is repeated at different values of the pressure drop between the inputs and outputs of the microchannel model.

During the experiment, in parallel, sensors are placed at the input and output points of the model to measure the difference in potentials. Using the "Keithley" potentiometer, the difference in potentials is set to an accuracy of ±0.001 mV [5].

4. Experimental study of the flow of Newtonian fluids in microchannels

After reaching a steady flow regime, at different values of the clearance openness (h) flow curves for tap water were generated—$Q = Q(\Delta P)$, dependence of the volumetric flow rate on the pressure drop, in the presence of atmospheric pressure at the outlet of the model. Based on the flow curves that were acquired, the hydraulic properties of the flow in microchannel are identified by revealing the dependences between the average share rate and the shear stress $\gamma = \gamma(\tau)$.

It is known that the volumetric flow rate of a liquid with a steady laminar flow between two stationary parallel plates is defined as $Q = bh^3\Delta P/12\,\mu L$, where, b, L, and h, respectively, are the width, length, and openness of a rectangular slit.

The values of γ and τ were determined as $\gamma = 6Q/bh^2$ and $\tau = \Delta Ph/2\,L$. Bingham fluid model was used to approximate the the curves $\gamma = \gamma(\tau)$. Rheological parameters of the liquid—the yield shear stress τ_0 and the apparent viscosity μ were also estimated on the basis of Bingham model.

For different values of openness h (15, 20, and 25 μm), the obtained curves $\gamma = \gamma(\tau)$ for water at a temperature 30°C were shown in **Figure 3**.

The flow curves associated with microchannels having an openness (h) greater than 25 μm are shown to be linear and corresponds to the Newtonian model. Water behaves like a non-Newtonian fluid with some yield shear stress τ_0 characteristic for the Bingham model, but the flow becomes nonlinear at values $h < h_{cr} = 25$ μm. As the slit's openness decreases, the water's non-Newtonian nature becomes more apparent, and the effect is most noticeable at the lowest value of h (10 μm) in the considered range.

Based on the equation of $\mu = bh^3\Delta P/12QL$, the fluid viscosity was calculated. The dependence of the apparent viscosity of water on the value of the openness (h) was shown in **Figure 4**. As can be seen, with a decrease in h, starting from the critical value ($h_{cr} = 25$ μm), the value of apparent viscosity increases nonlinearly, which reaches its maximum value at $h = 10$ μm. In the higher values of openness h, the change in the value of apparent viscosity is almost insignificant. However, in lower values, as the effect becomes stronger, the apparent viscosity is significantly increased up to 5 folds [5].

The electrokinetic factor plays an important role in the observed transformation of a Newtonian system into a non-Newtonian one, strengthening of rheological nonlinearity, and growth of hydraulic resistance in thin microchennels.

As previously mentioned, the key manifestation of electrokinetic effects occurs in thin microchannels when the transverse geometric dimensions of the microchannel commensurate with the dimensions of the EDL. As a result, ions are carried away from the outer diffuse part of the electric double layer at the water-metal surface boundary as water passes through the gap. This leads to the generation of a streaming potential, or a potential difference between the microchannel input and exit.

The streaming potential, in turn, causes ion transfer to reverse the flow of the liquid, which ultimately leads to the manifestation of additional resistance to movement and a corresponding increase in viscosity—a phenomenon called the electroviscosity effect, which is significantly reflected in the nature of the fluid flow.

Furthermore, it should be considered that tap water is actually a heterophase system that contains a large number of colloidal particles up to 100 nm in size in addition

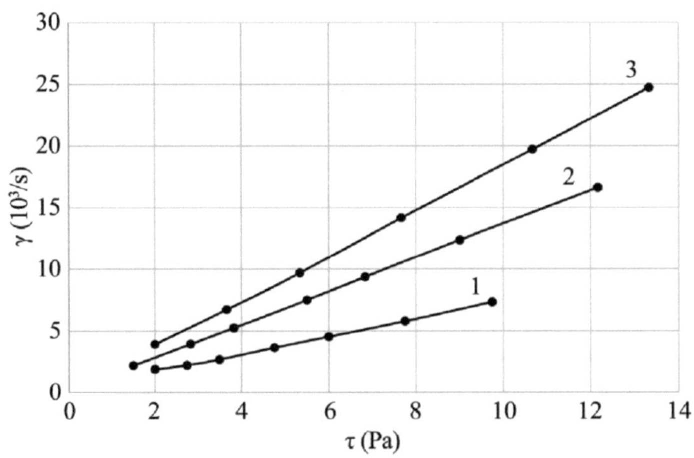

Figure 3.
$\gamma = \gamma(\tau)$ curves for water: 1–15, 2–20, and 3–25 μm.

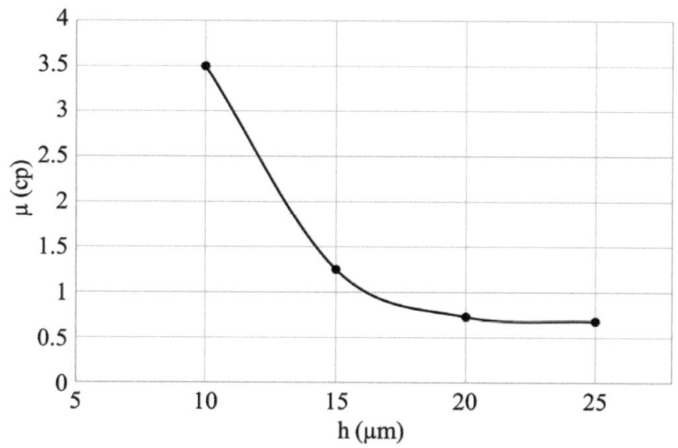

Figure 4.
The dependence of the apparent viscosity of water on the value of the openness h.

to suspended particles that are spread throughout the medium and range in size from several micrometers. When these particles come into contact with water, they all create EDL, which results in a complex composition of several local ion-electrostatic microfields dispersed across the dispersion medium's volume. An additional electrokinetic effect on the flow's character results from the zone of influence of these fields in the micro-slit becoming proportionate to the crack's openness.

Therefore, it is reasonable to assume that the hydraulic properties of the flow in a microchannel should be influenced by the system's electrokinetic potential and that these properties can be considerably altered by changing the system's electrokinetic potential.

Antistatic additives were chosen in order to control the flow's electrokinetic potential. The ND-12 reagent, which is often utilized as a demulsifier in oil-field settings, was employed as an antistatic agent.

Initially, an electrostatic cell was used to measure the electrode potential of water in order to establish the optimal concentration of additives. A platinum electrode was coaxially inserted into a stainless-steel cell to serve as the cell. The cell's body served as the second electrode. The electrode potential $\Delta\varphi$ was measured at various quantities after reagent's drops were applied to the water.

Figure 5 shows the dependence of the electrode potential $\Delta\varphi$ on the concentration (%) of the antistatic reagent. Drops of reagent were added to the water in the cell and the values of the electrode potential $\Delta\varphi$ were taken at different concentrations.

As can be seen, this dependence is not monotonous and the minimum potential is achieved at very small addition of the reagent. At the beginning, with an increase in concentration, the potential decreases; however, having reached a minimum, begins to increase to a certain maximum value, after which it practically remains unchanged at higher values of the additive.

It was found that the optimal concentration value at which the minimum potential is reached is approximately 0.006% (60 ppm). At this value of the additive, a multiple (10-fold) decrease in the potential is observed. A further addition of the antistatic is accompanied by an increase in $\Delta\varphi$ up to the certain value, which remains almost unchanged at concentrations greater than 0.04% (400 ppm). In further experiments, the flow curves for water were again plotted for the same microcracks, but with the presence of an antistatic additives.

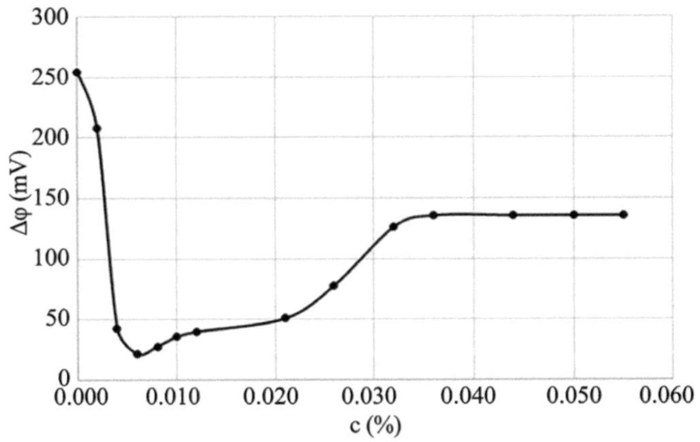

Figure 5.
The dependence of the electrode potential $\Delta\varphi$ on the concentration (%) of the antistatic reagent.

The $\gamma = \gamma(\tau)$ curves for water with an optimal concentration of antistatic reagent (60 ppm), for different values of h, at temperature 30°C were shown in **Figure 6**.

The following important conclusions can be drawn from the comparison of curves represented in **Figures 3** and **6**. The non-Newtonian character, manifested for water at h = 20 μm, practically, disappears in the presence of an additive and the flow becomes Newtonian. For slits with an openness h = 15 μm, a clear weakening of the non-Newtonian behavior is observed, with a significant decrease in the yield share stress, and, accordingly, in hydraulic resistance.

It should be noted that for a higher concentration (400 ppm), the effect is less pronounced. For comparison, **Figure 7** shows the dependences $\gamma = \gamma(\tau)$ for water in the presence and absence of additives during flow in cracks with an opening of 10 μm. As can be seen, antistatic additives lead to an essential weakening of the non-Newtonian character of water and reduction in hydraulic resistance.

The plotting of flow curves for each case was accompanied, simultaneously, by the measurement of the streaming potential $\Delta\varphi$. The values of $\Delta\varphi$, at $\Delta P = 10^5$ Pa, for

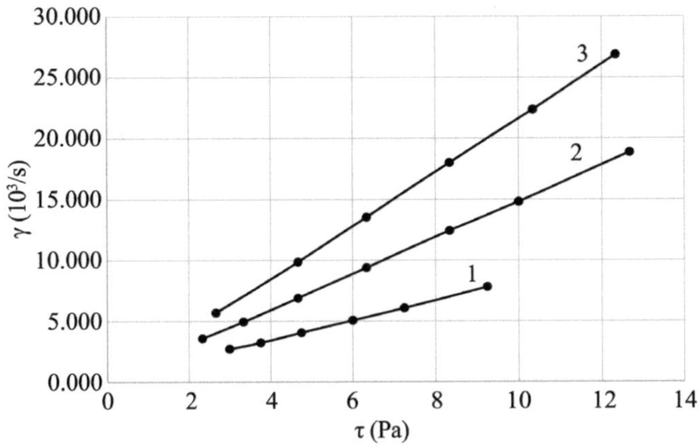

Figure 6.
$\gamma = \gamma(\tau)$ curves for water with an antistatic additive: 1–15; 2–20; 3–25 μm.

DOI: http://dx.doi.org/10.5772/intechopen.1008908

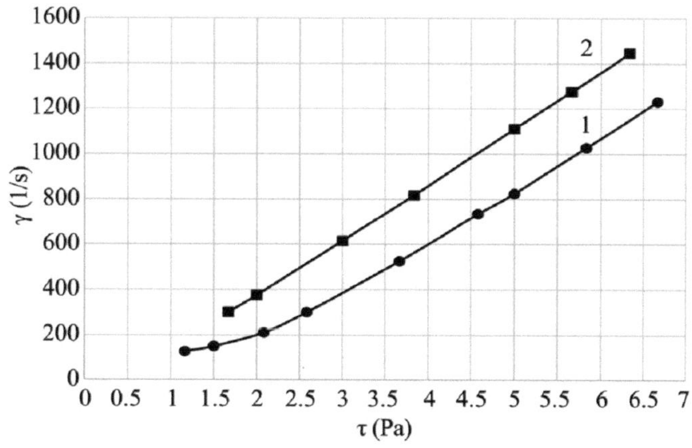

Figure 7.
$\gamma = \gamma(\tau)$ curves: 1 – water without an additive at 10 μm, 2 – water with an additive at 10 μm.

various values of h, in the absence and presence of additives of two concentrations, are shown in **Figure 8**.

As can be seen, in the flow of water containing additives of optimal concentration (60 ppm), the value of the streaming potential is significantly lower. It can also be seen that the effect is less pronounced for higher concentration (400 ppm).

The values of the yield shear stress τ_0 as a function of clearance h are displayed in **Figure 7**. As a function of openness h, τ_0 falls linearly in each of the three scenarios. It is evident that without additives, the water flow in the microchannel exhibits a non-Newtonian character, and the yield shear stress value rises as the gap narrows, peaking at the minimal clearance (10 μm). The inclusion of additives causes the critical value of the gap to shift towards smaller values of h. At a concentration of 60 ppm, a non-Newtonian character is already seen at h less than 20 μm, and this value increases as the slit openness decreases further. Furthermore, at a concentration of 400 ppm, critical thickness sets between line 1—water system with an additive concentration of 60 ppm—and 3—water

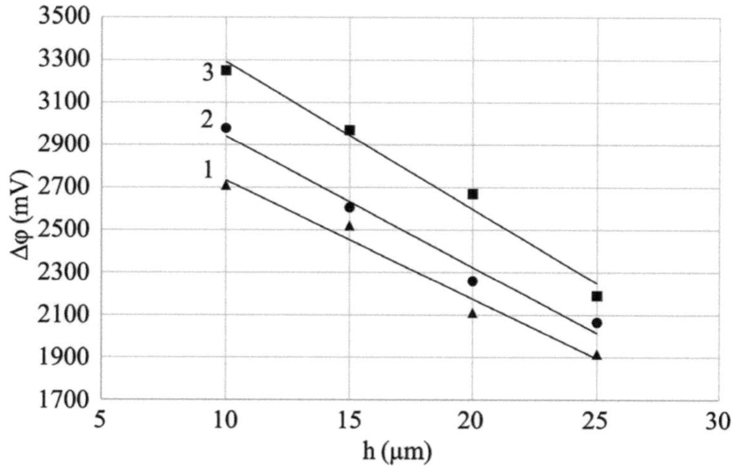

Figure 8.
The dependence of the $\Delta\varphi$ on the value of the openness h*: 1 – water with an additive of 60 ppm, 2 – water with an additive of 400 ppm, 3 – water without an additive.*

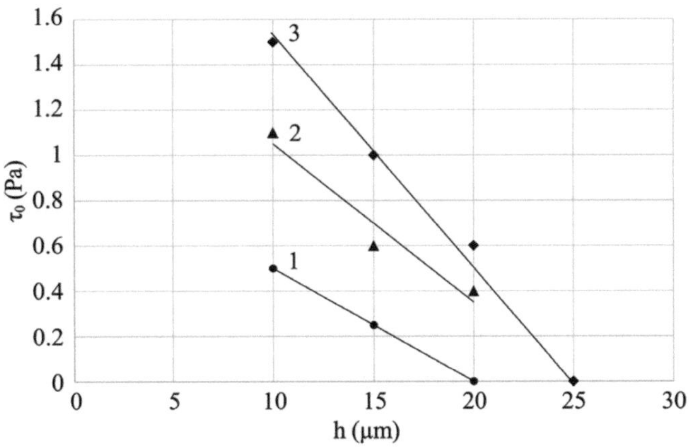

Figure 9.
The dependence of the yield shear stress τ_0 on the value of the openness h*: 1 – water with an additive of 60 ppm, 2 – water with an additive of 400 ppm, 3 – water without an additive.*

system without an additive. This trend complements each other with the potential differences of the corresponding water systems as shown in **Figure 9**.

The comparison demonstrates that the yield shear stress is significantly reduced when antistatic chemicals are included. Therefore, the value of the τ_0 decreases by more than three times for h = 10 μm.

To note that, the addition of antistatic reagent practically does not affect the change trend of the apparent viscosity as a function of openness h in the considered range and the curve shown in **Figure 3** remains unchanged.

Regardless of unchanged viscosity tendency, the flow parameters are significantly intensify. This intensification is due to decline in τ_0. This phenomenon has been, first time, discovered and named "electro-rheostress" effect.

5. Experimental study of the role of electrokinetic regulation on the intensity of transfer processes of hydraulic system in the microchannel model

One of the main factors affecting the effectiveness of the oil recovery process in low-permeable and microcrack oil fields is the transmissibility of porous and microcrack medium. Extensive number of scientific and practical works exists on increasing the oil recovery coefficient significantly by increasing the intensity of the transmission process in porous and microcrack environments by various methods. In this book chapter, the effect of reducing the electrokinetic potential of the fluid system on the intensity of the transfer processes of hydraulic system in the microcrack model was studied. The transfer processes of the microcrack model were evaluated by two different methods: 1 – analysis of pressure recovery curves (PRC) in different fluid systems and 2-analysis of the process of oil displacement by water.

5.1 Assessment of the effect of reducing electrokinetic potential on PRCs in water-microcrack systems

One of the methods to determine the formation parameters in low-permeable and microcrack oil fields is the pressure recovery method. On the basis of extensive laboratory works, it was established that by reducing the electrokinetic potential in water-microcrack systems, it is possible to regulate the intensity of the pressure recovery process and, as a result, significantly improve the flow parameters.

Laboratory studies were carried out with tap water. Antistatic additives were used to regulate the electrokinetic potential of fluid system.

Alternately, water without additives and with 0.006% antistatic additives were passed through the microcrack model. The process was repeated at four different openness – 10, 15, 20, and 25 μm-and PRCs were constructed (**Figure 10**).

Figure 11 represents PRCs at four different openness for water at the optimum concentration of the additive. From PRCs, it is clear that during the flow of water through the microcrack model, starting from h_{kr}, the pressure recovery time and relaxation time increase with the decrease of openness h. Pressure recovery time at the opening of h = 25 μm was 50 s, while it is 250 s at the openness of 10 μm.

At the same time, a complete restoration of pressure at the entry and exit points of the microchannel is recorded at the values of $h \geq h_{kr}$. However, as h decreases, the difference in pressures at the entry and exit points is manifested, reaching a peak value at the smallest value of h in the considered range (h = 10 μm). This process can be explained by the manifestation of electrokinetic factors, as well as the EDL effect at the openness below h_{kr} and its intensification as h decreases. Consequently, a Newtonian fluid makes a transition to a non-Newtonian fluid, and as h decreases, the non-Newtonian properties of the fluid flow become sharper.

From the comparative analysis of **Figures 12** and **13**, it is clear that under the influence of antistatic additives, the intensity of pressure recovery during the flow of water in the microcrack model has been significantly improved. Thus, at the openness of h = 10 μm, the recovery of the flow pressure of water was recorded at 250 s. With the effect of optimal concentration of antistatic additive, this value decreased to 150 s. In the considered range, similar intensification in the pressure recovery process was recorded at other values of openness h.

Under the influence of the antistatic additive, a decrease in the pressure difference manifested between the entry and exit points of the model was also recorded at values below critical openness. At the smallest value of h = 10 μm in the considered range, the pressure difference through the model was decreased from 10^5 Pa to $0.3*10^5$ Pa

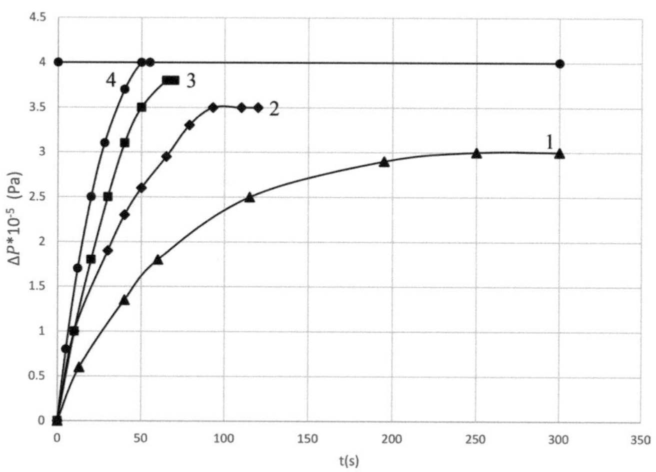

Figure 10.
Pressure recovery curves during water flow at four different openness: 1–10, 2–15, 3–20, 4–25 μm.

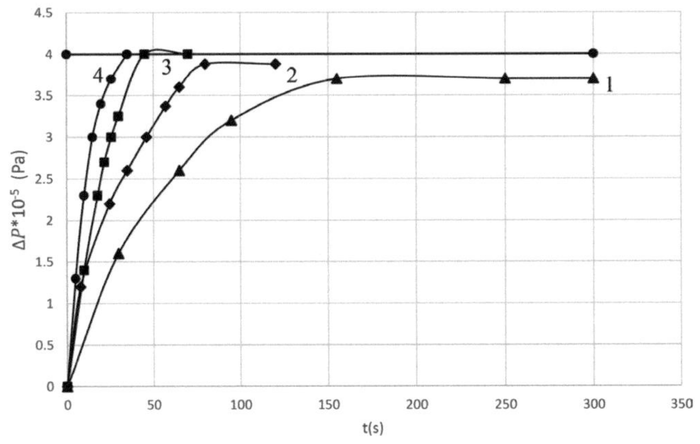

Figure 11.
PRC during flow of water with 0.006% antistatic reagent at four different openness: 1–10, 2–15, 3–20, 4–25 μm.

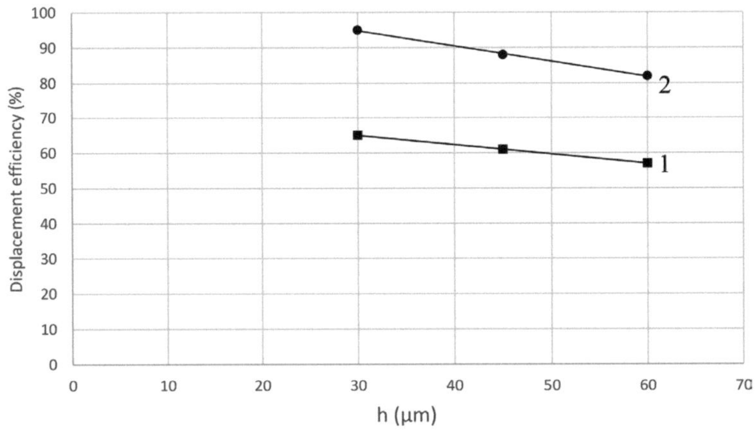

Figure 12.
Displacement efficiency as a function of openness h *at* $\Delta P = 2*10^5$ *Pa: 1 – in the absence of antistatic additive and 2 – in the presence of antistatic additive.*

with the effect of optimal concentration of the antistatic additive. While $0.2*10^5$ Pa pressure difference appears through the model at the openness of $h = 20$ μm, with the effect of antistatic reagent, a complete restoration of pressure was provided.

Table 1 shows the pressure differences between the inlet and exit points of the model in the case of water in the absence and presence of antistatic additive for each openness h.

The recorded results are also consistent with the tendency of the stream potential to change under the influence of the antistatic additive. In the considered range, a decrease in the flow potential was recorded under the influence of the optimal concentration of the antistatic additive, which results in a decrease in the pressure recovery time at the corresponding values of h.

In **Figure 14**, in the considered range, a graph of the dependence of the pressure recovery time h for water without additives and water with antistatic additives of 0.006% is given. A decrease in pressure recovery time was observed in all values of h under the influence of the additive. The effect of the antistatic additive, on the other hand, increases as h decreases and reaches its maximum limit at the minimum value of $h = 10$ μm in the considered range.

DOI: http://dx.doi.org/10.5772/intechopen.1008908

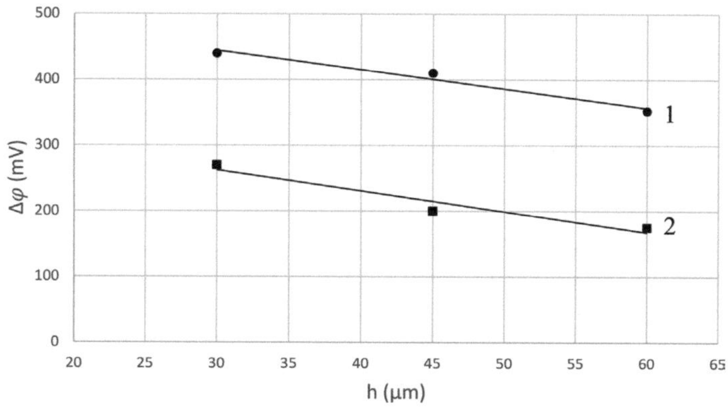

Figure 13.
The dependence of stream potential on the openness h *of the microchannel model: 1 – water without antistatic agent, 2 – water with an antistatic agent.*

Openness h (μm)	ΔP in water without antistatic additive (Pa)	ΔP in water with 0.006% antistatic additive (Pa)
10	1	0.3
15	0.5	0.12
20	0.2	0
25	0	0

Table 1.
The pressure difference arising at the inlet and outlet of the model after stabilizing the water flow.

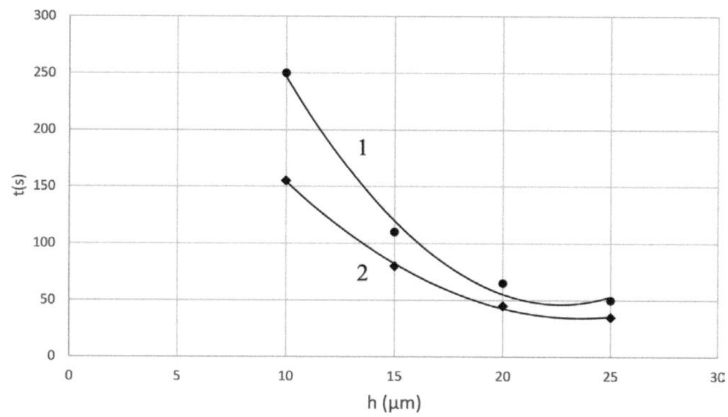

Figure 14.
Dependence of pressure recovery time on the openness h*: 1-water without antistatic additive, 2–water with an antistatic additive of 0.006%.*

For all values of h, in the considered interval, relaxation times were set based on the function ln((Pin-pout)/(P-pout)). At high values of h, the relaxation time is zero, since the hydraulic resistance to fluid flow is minimal. As h decreases, the electrokinetic effects intensify and additional hydraulic resistance to flow is created. Consequently, the relaxation time is manifested and increases as h decreases. The relaxation time obtained from the ln((Pin-Pout)/(P-pout) function almost coincides with the pressure recovery time. For water without additives and water with antistatic agent of 0.006%, the relaxation times for h in the considered interval are shown in **Table 2**.

5.2 Electrokinetic regulation of the process of oil displacement by water in microcracks

In oil and gas fields, especially in low-permeable and microcracked hydrocarbon formations, when oil is displaced by water, remaining a large amount of oil residuals underground is a serious problem, which still requires an effective solution. Large scale scientific and practical works have been carried out in order to increase the mobility of formation fluids in low-permeable and microcracked oil and gas fields, as well as to improve the efficiency of the displacement process and reduce the amount of hydrocarbons remaining underground as residual oil and gas. However, in current studies, the control of the process through electrokinetic regulation of the rheophysical properties of the displacing fluid has not been considered.

In the process of oil displacement with water, especially in microcracks, displacing properties of water are determined on the basis of surface and capillary phenomena resulting from surface tension at the oil-water contacts. As is known, surface tension is the manifestation of intermolecular forces, which are essentially electrostatic. Based on this, it can be concluded that the ability of water to displace the oil can be increased by adjusting the electrostatic potential of the displacing water in various ways [8].

Extensive laboratory works were carried out in order to experimentally study the effect of electrokinetic regulation of the rheophysical properties of fluid systems on the process of oil displacement by water in the microchannel model. The experiments were carried out at a constant temperature—30°C. In order to ensure the isothermal process, the model was placed in a thermostat and connected to an ultrathermostat to regulate the temperature of the displacing fluid.

In the experiments, Newtonian oil with viscosity of 48cp was used as a displaced liquid, and water was used as a displacing liquid. The microchannel model was repeatedly vacuumed and saturated with Newtonian oil. The weight of the model was measured before and after saturation with oil, and the amount of oil injected into the model was determined based on the difference in weights. The model is considered fully saturated with liquid after passing through the model an amount of fluid approximately 20–30 times the volume of the slit of the model. With this method, saturation of the model with liquid is ensured 97–98%. During the experiments, the input pressure of the model was changed in the range of $2*10^5$–$5*10^5$ Pa. The outlet pressure, however, was kept constant at 10^5 Pa.

Openness h (μm)	t in water without antistatic additive (san.)	t in water with 0.006% antistatic additive (san.)
10	204	143
15	95	54
20	59	41
25	52	36

Table 2.
Relaxation times for each openness h *for water without additive and water with antistatic additive of 0.006%.*

Initially, oil was displaced with water and displacement efficiency were estimated under different pressure drops. In subsequent experiments, the electrokinetic potential of the displacing liquid–water as well as its rheophysical properties were regulated by adding an antistatic additive with optimal concentration, and its effect on the displacement efficiency was studied. The experiments were repeated at three different openness of the microchannel model—30, 45, 60 μm. The volume of displaced oil was calculated by gravimetric method. The final displacement efficiency is determined by the ratio of the displaced oil to the amount of the initial oil injected into the model.

Figure 15 shows the dependence of the displacement efficiency of Newtonian oil by water on pressure drop across the model at three different openness—30, 45, and 60 μm. At all considered values of h, the displacement efficiency of oil by water decreases as the inlet pressure of the model increases. This can be explained by the intensification of the sliding effect of the fluid, which compresses the flow rate in excess, in relation to the liquid being compressed. And as the openness of the microchannel model decreases, there is an increase in the displacement efficiency.

In further experiments, the electrokinetic potential of the displacing fluid was reduced using the optimal concentration of the ND-12 antistatic reagent. **Figure 16** shows the curves of dependence of the displacement efficiency of oil displaced by water with optimal concentration of antistatic reagent on the pressure drop at three different values of openness h. In all the considered values of h, reducing the electrokinetic potential of the displacing fluid under the influence of an antistatic additive results in a significant increase in the displacement efficiency (up to 30%).

For comparison, **Figure 12** shows the dependence of the displacement efficiency of oil by water without and with antistatic additives under the pressure drop of $\Delta P = 2{*}10^5$ Pa. In all the considered values of h, it is possible to clearly see a significant intensification in displacement process with the effect of an antistatic reagent. At the same time, in both fluid systems, with an increase in h, a decrease in the displacement efficiency was recorded.

Parallel to the displacement process, the stream potential was also measured between the input and output of the system. **Figure 13** shows the dependence of the stream potential on the openness of h under the pressure drop of $\Delta P = 2{*}10^5$ Pa in the displacement process of oil by water with the absence and presence of antistatic

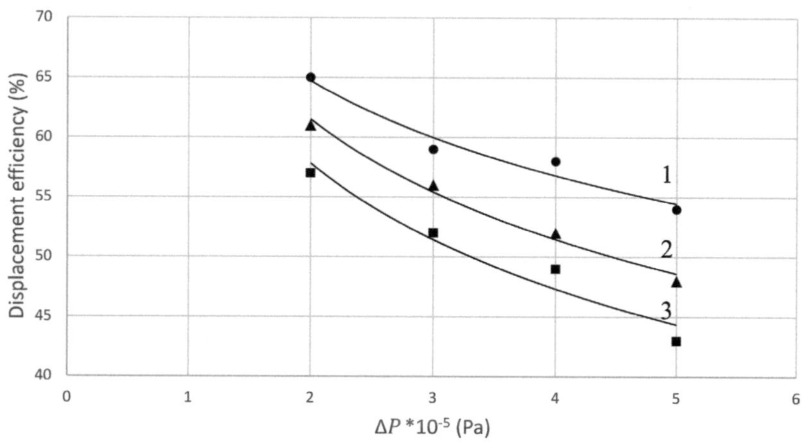

Figure 15.
Displacement efficiency as a function of ΔP *in the absence of antistatic additive, 1–30, 2–45, 3– 60* μm.

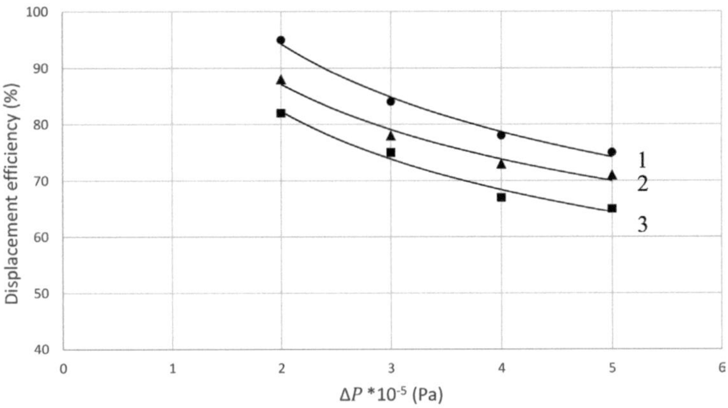

Figure 16.
Displacement efficiency as a function of ΔP in the presernce of antistatic additive: 1–30, 2–45 ,3 – 60 μm.

additives. At the openness of h = 30 μm, the streaming potential in the displacement process of oil with water was 440 mV, but under the influence of the antistatic reagent, it was decreased to 270 mV. At the values of 45 and 60 μm of openness h, similar decreases in flow potential were also recorded. The results obtained reaffirm the significant effect of reducing the electrostatic potential of the displacing fluid on the effectiveness of the displacement process.

6. Conclusion

The influence of electrokinetic factors on the rheophysical properties of the fluid flow in microchannels has been studied on the basis of microchannel model. It has been established that in the filtration process of viscous liquids in microchannel, the electrokinetic potential of the system has a special role in the manifestation of non-Newtonian properties, accompanied by a violation of Darcy's law. Thus, reducing the electrokinetic potential of the system using chemical methods significantly weakens the non-Newtonian character of the flow and, as a result, improves the filtration properties of the microchannel.

Small and ultrasmall concentrations of antistatic additives significantly reduce the electrostatic potential of the fluid system. The curves of dependence of the electrode potential ($\Delta\varphi$) of liquids (water, oil) on the concentration of antistatic additives in the system are nonlinear, and at an ultrasmall concentration of the antistatic additive, the electrostatic potential decreases to the minimum value of $\Delta\varphi$.

On the basis of $\gamma = \gamma(\tau)$ curves of fluid systems in the presence and absence of the antistatic additives, based on the Bingham model, the rheological parameters of the fluid flow were evaluated in different openness. The optimal concentration of the antistatic reagent strongly reduces τ_0 of the system, by reducing the electrokinetic potential of the fluid flow, and, consequently, significantly improves the filtration properties of microchannel. At smaller openness, the effect is manifested more strongly.

By reducing the electrokinetic potential of the liquid system under the influence of the optimal concentration of the antistatic reagent, the critical value of the openness decreases, and as a result, the manifestation of non-Newtonian properties is delayed and significantly weakened. The blocking of fluid flow under the influence of the EDL effect also occurs at smaller openness. As an example, in the water-microcrack model, the critical openness of the slit was reduced from 25 to 20 μm.

Reducing the electrokinetic potential of the fluid flow with the optimal concentration of antistatic additives also has a significant effect on the intensity of non-stationary processes. In the absence and presence of antistatic additives, from the comparative analysis of pressure recovery curves in microchannels of different sizes, it is clear that with a decrease in the electrical potential, the rheophysical transmission process in the system intensifies and, as a result, the pressure recovery time in the hydraulic system is significantly reduced.

It has been established that, in the process of oil displacement from microchannels, the presence of an ultrasmall amount of antistatic additive in water and, accordingly, the reduction of electrokinetic potential of the system significantly increased the displacement efficiency of water. The obtained results give a reason to believe that water containing ultrasmall optimal concentration of antistatic reagent can be used as a highly effective displacing agent in oil displacement process from microchannels.

Reducing the electrokinetic potential of the fluid system with antistatic additives has a significant impact on the intensity and effectiveness of the oil displacement process in the microchannel model. The results of experiments on oil displacement by water in the absence and presence of antistatic additives under different pressure drops show that the displacement efficiency increases by 30% with the effect of optimal concentration of antistatic additives.

Considering the electrokinetic effects, the study of the rheophysical properties of the fluid flow in microchannels and the improvement of the filtration properties of microchannels through electrokinetic regulation of the electrostatic potential of the hydraulic system can be used to significantly increase the economic efficiency of development and exploration of low-permeable and microcracked oil fields.

Author details

Fuad Veliev* and Aida Aslanova
Baku Higher Oil School, Baku, Azerbaijan

*Address all correspondence to: fuadveliev9@gmail.com

References

[1] Gandossi L, Bocin-Dumitriu A, Spisto A. International Developments in the Field of Unconventional Gas and Oil Extraction. Luxembourg: Publications Office of the European Union; 2017. pp. 1-62. DOI: 10.2760/372964

[2] Rafiqul MI. Basement reservoirs. In: Sustainable Oil and Gas Development Series, Reservoir Development. Gulf Professional Publishing; 2022. pp. 533-607. Available from: https://www.amazon.com/Reservoir-Development-Sustainable-Oil-Gas/dp/0128200537. DOI: 10.1016/B978-0-12-820053-7.00010-X

[3] Muggeridge A, Cockin A, Webb K, Frampton H, Collins I, Moulds T, et al. Recovery rates, enhanced oil recovery and technological limits. Philosophical Transactions of the Royal Society A. 2014;**372**:1-24. DOI: 10.1098/rsta.2012.0320

[4] Veliyev FH, Aslanova AR. Electrokinetic Regulation of Non-Newtonian Character of Water Flow in Microcrack Model. Almatı: Нефтехимия; 2023. pp. 136-143

[5] Veliyev FH, Aslanova AR. A rheophysical study of the non-Newtonian behavoir of water flow in thin channels. Estonia, Eureka: Physics and Engineering. 2024;**N1**:11-17. DOI: 10.21303/2461-4262.2024.003050

[6] Hajmohammadi MR, Bahrami M, Ahmadian-Elmi M. Thermal performance improvement of microchannel heat sinks by utilizing variable cross-section microchannels filled with porous media. International Communications in Heat and Mass Transfer. 2021;**126**:1-14. DOI: 10.1016/j.icheatmasstransfer.2021.105360

[7] Veliyev FH, Aslanova AR. Non-Newtonian characteristics of water flow in microchannels. Indian Journal of Science and Technology. 2023;**16**:4605-4611. DOI: 10.17485/IJST/v16i47.2691

[8] Veliyev FH, Aslanova AR. Electrokinetic Regulation of the Process of Oil Displacement by Water in Microcracks. Baki: Azerbaijan Oil Industry; 2023. pp. 18-21. DOI: 10.37474/0365-8554/2023-12-18-21

[9] Veliyev FH, Aslanova AR. Rheophysical characteristics of hydrocarbon flow in microcracks, Las-Vegas. International Journal on Engineering, Science and Technology. 2024;**2**(6):163-170

[10] Veliyev FH, Aslanova AR. Electrokinetic regulation of flow in microcrack reservoirs. In: Osaka: 26th IUPAC International Conference on Chemical Thermodynamics, Abstract Book. Japan: Japan Society of Calorimetry and Thermal Analysis Science Council; 2023. p. 103

[11] Diwu P, Liu T, You Z, Jiang B, Zhou J. Effect of low velocity non-Darcy flow on pressure response in shale and tight oil reservoirs. Fuel. 2018;**216**:398-406

[12] Chen Z, Liao X, Zhao X, Lyu S, Zhu L. A comprehensive productivity equation for multiple fractured vertical wells with non-linear effects under steady-state flow. Journal of Petroleum Science and Engineering. 2017;**149**:9-24

[13] Cheng Z, Ning Z, Zhang W, Ke S. Theoretical investigation of electroviscous flows in hydrophilic slit nanopores: Effects of ion concentration and pore size. Physics of Fluids. 2020;**32**(2):022005

[14] Wu J et al. Experimental study of nonlinear flow in micropores under low pressure gradient. Transport in Porous Media. 2017;**119**:247-265

[15] Wu J et al. Flow of Newtonian fluids with different polarity in micro scale. Chinese Science Bulletin. 2017;**62**:2988-2996

[16] Bahrami M, Tamayol A. Laminar flow in microchannels with noncircular cross section. Journal of Fluids Engineering. 2010;**132**:1-10

[17] Bahrami M, Culham JR, Yovanovich M. A novel solution for pressure drop in singly connected microchannels of arbitrary cross-section. International Journal of Heat and Mass Transfer. 2007;**50**:2492-2502

[18] Shahsavari S, Tamayol A, Kjeang E, Bahrami M. Convective heat transfer in microchannels of noncircular cross sections: An analytical approach. Journal of Heat Transfer. 2012;**134**:091701

[19] Sadeghi E, Djilali N, Bahrami M. Estimation of Nusselt number in microchannels of arbitrary cross-section with constant axial heat flux. In: Proceedings of the Sixth International ASME Conference on Nanochannels, Microchannels and Minichannels ICNMM, June 23-25, Darmstadt, Germany. Germany: ASME; 2008

[20] Akbari M, Sinton D, Bahrami M. Pressure drop in rectangular microchannels as compared with theory based on rbitrary cross section. Journal of Fluids Engineering. 2009;**131**:1-8

[21] Bahrami M, Tamayol A, Taheri P. Slip-flow pressure drop in microchannels of general cross section. Journal of Fluids Engineering. 2009;**131**:031201

[22] Bahrami M, Yovanovich M, Culham J. Pressure drop of fully-developed, laminar flow in microchannels of arbitrary cross-section. Journals of Fluid Engineering. 2005;**128**:1-10

[23] Loeb LB. Static Electrification. Berlin Heidelberg: Springer; 1958

[24] Benjamin R, Emily M, Parshotam S, Emma L, Franz G, Julianne M. Water structure in the electrical double layer and the contributions to the total interfacial potential at different surface charge densities. Journal of the American Chemical Society. 2022;**144**(36):16338-16349. DOI: 10.26434/chemrxiv-2022-p5fwp-v2

[25] André H, Dourado B. Electric double layer: The good, the bad, and the beauty. Electrochem. 2022;**3**:789-808. DOI: 10.3390/electrochem3040052

[26] Zhang L, Chenkun L, Huang J. A beginners' guide to modelling of electric double layer under equilibrium, nonequilibrium and AC conditions. Journal of Electrochemistry. 2022;**28**:1-30

[27] Mujtaba I, Ali R, Muhammad I, Asif M. Carbon Nanocomposite-Based SCs as Wearable Energy Storage. Springer International Publishing; 2022. Available from: https://www.springerprofessional.de/en/carbon-nanocomposite-based-scs-as-wearable-energy-storage/22166568

[28] Maltsev V, Stoke M, Tsoutsanis P. High-order methods for diffuse-interface models in compressible multi-medium flows: A review. Physics of Fluids. 2022;**34**:1-34

[29] Mahmoud K, Dominik P. Structure of the electrical double layer revisited: Electrode capacitance in aqueous solutions. Langmuir. 2020;**36**:36-47

[30] Budin M. Electrophoresis Fundamentals. De Gruyter; 2022. p. 522. Available from: https://www.degruyter.com/document/doi/10.1515/9783110761641/html?lang=en&srsltid=AfmBOopBgnSIgPyujW1GKBUMbWzql7cwCecYdCCW16in0MVJnFB2ZkVY

[31] Gao Y, Wang C, Gong Z, Li Z. A mini review on electroosmotic phenomena in porous media. Energy Storage and Conversion. 2024;**2**:1-14. DOI: 10.59400/esc.v2i1.480

Chapter 3

Rockfill Hydraulics: Depths and Discharges in Single Gabion Weirs

Harry Edmar Schulz and André Luiz Simões

Abstract

A theoretical treatment of the rockfill hydraulics for the discharge of through flows in gabions is presented. Classical procedures are used and reviewed, and new results for flow depths and discharges are obtained. The equations for conservation of mass and momentum are taken in their integral form, together with the Bernoulli equation for conservation of energy (also known as "Bernoulli with losses"). Further, the usual classification of flows for gabions, including front through flows, upper through flows, and overflows, is complemented with full and partial front and upper through flows. In rockfills, the porous matrix in general has a large-scale porosity, which allows the occurrence of turbulent flows in their open spaces. Thus, laminar and turbulent flows were considered, and the new results are of simple application. The comparison of predicted depths and discharges for through flows with the results of the literature shows that the present propositions are adequate, and compose a useful complementation to the study of flows through gravels in rockfill hydraulics.

Keywords: gabions, hydraulics of gravel media, physics of flows through rockfill, flow depths and discharges in large erratic interspaces, through flow discharges

1. Introduction

Registers show that since more than 2000 years the mankind uses grid cages filled with stones or gravel for several uses, like the control of river erosion in ancient Egypt. The characteristics of form and production of such cages did not change much until the nineteenth century, when the industrial production of the so-called gabions began in the last decade of that century [1]. The name "gabions" is today a common designation of the grid cages filled with stones, while the name "rockfill" is of more general use, for works that may or not use gabions (grid cages) in the work site.

Due to its easy application, for example, in relatively small valleys and creeks, gabions have been used with a growing frequency to build dams or dam weirs as a low-cost solution for several flow problems, like the mitigation of erosion and flash floods [2]. In the present days, considering the ecological aspects, gabion weirs are taken as better suited to avoid the negative impacts in relation to their older solid-impermeable similars. The free-passing of substances and microorganisms relevant to the surrounding environment and the enhanced absorption of oxygen by the water are among the mentioned examples [1, 3, 4].

Being adequate to partially retain water upstream of their geographic location, and being economically attractive, studies related to calculate the discharges enabled by gabion weirs became relevant. In this sense, several researchers around the globe are furnishing equations to predict such discharges, which depend on the intrinsic characteristics of the flow and of the gabions themselves.

Taking a one-step gabion, different flow conditions can be observed, sometimes also called flow regimes: (1) front Through flow: The upstream water level does not surpasses the height of the gabion, so that the water discharge penetrates the gabion through the vertical front surface, and passes totally through the spaces formed between the rocks or gravels; (2) upper through flow: The upstream water level surpasses the height of the gabion and penetrates the gabion through the horizontal upper surface. This condition may occur with or without overflow, and when there is no overflow, it may take the whole horizontal surface or only part of it; (3) overflow: the exceeding water discharge that does not passes through the rocks or gravel, but behaves like passing over a solid impermeable weir.

There are several slightly modified denominations for the described conditions or regimes, like through flow, transient flow, and free overflow (as in [5]), but the present names describe better what is happening between fluid and gabion. The different parcels of discharge of the three flow conditions depend on the geometrical characteristics of the rocks or gravel that compose the material of the gabion, and of the total applied discharge.

Although being used throughout a millenary scale of time, objective calculations considering gabions are relatively recent (sparse studies along the last century). They involve various aspects, like the discharges that pass through the gabion, the free overflow discharge, and the energy dissipation rate. Historically, these studies came in the sequence of more detailed studies on impermeable free overflows or water drops, which were initiated around the half of the twentieth century and focused more on the energy dissipated by water drops. Although directed to solid impermeable steps, they can also orientate studies for permeable steps.

1.1 Impermeable one-step studies

Gill [6] initiated his description of vertical impermeable water drops citing previous studies conducted in the 1940–1950's, mentioning that his quantification of flow characteristics and energy dissipation led to better results than his precedents. The mentioned cause was that the level of assumptions in his study was not so drastic as those of the previous works. Gill [6] furnished results for the depth of the pool formed behind the falling water jet and suggested the so-called Bélanger Eq. (1) to provide right values of the ratio between sequent depths of hydraulics jumps. However, although not evidenced at that time, this equation does not consider bottom shear stresses, and other solutions may also be used, like Eq. (2), in which the bottom shear stresses are already computed [7–9].

$$\frac{y_2}{y_1} = \frac{\sqrt{1 + 8F_1^2} - 1}{2} \tag{1}$$

$$\frac{y_2}{y_1} = \phi_1 F_1^2 + \sqrt{\left(\phi_1 F_1^2\right)^2 + \phi_2 F_1^2} \tag{2}$$

F_1 is the supercritical Froude number, $y_{1,\,2}$ are the sequent depths of a hydraulic jump, and $\phi_{1,\,2}$ are coefficients that depend on the characteristics of the structure in which the jump is forming, like its roughness and flow characteristics. They resemble the friction factor in tube flows, which depend on flow (Reynolds number) and geometry factors. The functions for $\phi_{1,\,2}$ need further analyses of experimental data.

Rajaratnam and Chamani [10] focused on the energy dissipation of impermeable water drops and the pool depth behind the falling water jet. Those authors presented a resume of the previous studies, also describing the assumptions of the 1940–1950's studies and the improvements suggested by Gill [6]. Instead of theoretical proposals for the energy loss, an empirical equation was proposed, showing good correlation with own measured data and with data from the literature. The mentioned authors considered a single water drop in their studies, not including the length of the step upstream of the drop. In the sequence, Wu and Rajaratnam [11] studied the characteristics of subcritical approaching flows over an impermeable square step, considering the tail height in relation to the approaching height. Two flow regimes were observable in the experiments, described as (1) impinging jet and (2) surface flow regimes. It was concluded that the conditions of the hydraulic jumps may vary between three configurations: (1) repelled jump, (2) the jump linked to the basis of the drop, and (3) the submerged jump. The authors [11] controlled the characteristics of their experimental study by the downstream water depth, restricting the possible variations to the two mentioned flow regimes. Instead of subcritical flows, Chamani and Beirami [12] presented a study for supercritical flows at impermeable drops. The authors observed that a sequence of specific discharges that increased the Froude number implied in decreasing pool depths, downstream depths, and relative energy losses. Their experimental study enabled them to develop an empirical equation for supercritical energy losses.

Chamani et al. [13] then advanced in the detailed study of impermeable water drops, refining the understanding of the energy dissipation and searching for a better generalization of the formulation. The authors presented the derivation of an equation for the relative energy dissipation in which the height of the drop, the critical depth, and the depth at the basis of the flow were considered.

A study on impermeable steps that approach a bit more the condition of the usually studied gabions steps was presented by Azimi and Rajaratnam [14], in which steps with finite lengths were considered. The authors worked with the usual basic equation for discharges through rectangular weirs, using extensive data of the literature in their analyses. As known, the quantification of the discharges depends on the use of adequate discharge coefficients, and the authors showed that a classification of flow/weirs into four classes led to good empirical correlations. The proposed classes are long-crested, broad-crested, short-crested, and sharp-crested weirs. Further, Amin et al. [15] concentrated studies on flows over sharp-crested weirs, for the conditions of free and submerged flows, focusing once more on the energy dissipation. Six empirical equations for energy dissipation were provided for different flow conditions.

As can be observed in the previous paragraphs, the studies of impermeable water drops were more centered on the energy dissipated in such structures. Like mentioned, the study of finite step weirs is a bit more related to this study, providing correlations for coefficients to be used in the calculations of the discharge passing over the steps. However, this study is mainly focused on the discharges passing through the gabion step, and not over it.

1.2 Permeable multisteps studies: Gabion-stepped weirs

Dust and Wohl [16] studied flows in natural steps crests, and presented some situations that resemble the condition of impermeable steps, together with additional conditions that resemble gabion flows. Among their descriptions and definitions, the condition of interstitial flow is illustrated (flow occurring in the spaces formed between rocks that partially block a water course), together with the weir flow and the oscillatory flow. Considering the discharge coefficient C^* for the discharge Eq. (3), the graphical information [16] provides a nondimensional value of C^* between 0.25 and 0.30 for the interstitial flow.

$$Q = C^* B\sqrt{g}h^{3/2} \tag{3}$$

Q is the discharge, B is the width of the channel, g is the acceleration of gravity, and h is the water depth. The flows through gabions may also be viewed as "horizontal flows through packed columns." This enables to use analytical procedures described in the literature of packed columns and of porous media. Among the procedures are the use of equations for laminar head losses (Darcy equation and Kozeny-Carman equation) and for combined laminar and turbulent head losses (Forchheimer equation and Ergun equation), as shown by Kozeny [17] and Carman [18, 19].

Considering the constructive aspects of structures of gabions, the report "Les Ouvrages en Gabions – Techniques Rurales en Afrique" [20] presents an impressive selection of practical solutions for hydraulic and geotechnical questions used in rural areas of the former République de Haute-Volta, in Africa, today named Burkina Faso. The descriptions of the confection of gabions with several geometrical forms: their use as dams, as retaining walls, weirs, channel protections, vertical walls, stepped walls, inclined walls, main or auxiliary building elements, among others, are still reproduced with little descriptive modifications in more recent publications on the topic. A further reference that presents a broad variety of applications of gabions, and of solutions for flows through and over them (for a single gabion, for stepped weirs, for supporting walls and other protective applications) is Stephenson [2]. The chapters of his book "Rockfill in Hydraulic Engineering" may be considered classical works that still today help to direct studies in this field. Like Ref. [20], the work [2] is concentrated in the African continent, in this case in South Africa.

The literature on the use of gabions is becoming "denser" since the studies by Peyras et al. [21, 22], with the contributions increasing more intensively in the last two decades. Many studies were focused on stepped channels, chutes, and weirs, presenting valuable experimental results for adequately chosen flows and geometrical conditions, and providing corresponding empirical equations for different variables. The benefits obtained from these experimental studies are evident. Even theoretical and conceptual developments need the experimental basis to adequately justify their results. In this sense, this study is also among those that rely on aspects previously described by experimental and theoretical studies of both stepped gabion structures and one step weirs.

Peyras et al. [21, 22] studied gabion-stepped chutes and focused on the energy dissipation provided by these structures. They provided a methodology to design dissipation basins based on the Bélanger equation and on the empirical length L of the basin given by $L = 6Y_2$, where Y_2 is the subcritical conjugate depth of the hydraulic jump formed in the basin.

As general information about the use of gabions, in 2001 the Land and Water Development Division of the Food and Agriculture Organization of the United Nations published a report containing constructive aspects of small dams, considering earth dams and gabions [1]. The report presented several aspects of the installation of small dams, like the different types, the characteristics of needed hydrological and geological studies, with the description and classification of soils and rocks. Further, the design of gabion spillways and embankments was considered, together with topics for the construction and quality control of earth and gabion materials, and the construction, control, and maintenance of small dams. The literature of application of gabions provides several examples of application, like the erosion control, stabilization using vegetated gabions [23], and multiple hydraulic structures [3], among others.

A detailed study of stepped gabion weirs was presented by Chinnarasri et al. [24], whose results were unified through nondimensional parameters obtained from adequate dimensional analyses. Data were collected for several channel slopes; discharges; geometrical characteristics of the gabions; diameters and roundness of the used stones; and for the nappe, transition, and skimming flow regimes. Results for energy dissipation, velocities, and pressure on the steps were presented, together with empirical nondimensional equations.

Khatibi et al. [25] searched specifically the energy dissipation in stepped gabion weirs. The authors studied skimming and nappe flow regimes, different slopes and porosity, and flow characteristics quantified through the Reynolds and the Froude numbers. To evidence the transition between nappe and skimming flows, the data of energy dissipation were plotted against GN, a version of the drop number of Rand [26], named Gabion Number by Chinnarasri et al. [25]. It was shown that the nappe flow regime occurs for $GN < 0.01$, the skimming flow regime occurs for $GN > 0.02$, and the transition regime occurs for $0.01 < GN < 0.02$.

Also focused in energy dissipation, considering still air entrainment, Refs. [27, 28] presented measured velocity profiles, together with air concentration profiles and turbulence properties over stepped flumes. Wűthrich and Chanson [27] presented very detailed images of the experimental conditions for flat (impermeable) surfaces and gabion (permeable) surfaces, so that comparisons between the characteristics of both situations could be made. The authors provided also results for mass transfer, that is, the absorption of dissolved oxygen by the flowing water, a matter also studied by Minho et al. [29] for impermeable stepped flumes, who provided practical empirical equations that can be used for comparisons with flows over gabion steps. Velázquez-Luna and Ventura-Ramos [30], on the other hand, conducted experiments to verify the capacity of retaining sediments for gabions having stones with different diameters and forms. The authors also observed that the slope of the channel influences decisively on the retention characteristics of gabion weirs.

Reeve et al. [31] developed a numerical study to evaluate the performance of gabion stepped spillways using Reynolds Averaged Navier-Stokes (RANS) equations. Several geometries of gabion steps, channel slopes, and porosity (gravel sizes) were tested, calculating the energy dissipation and the inception point. They presented their conclusions commenting the reduction of the non-aerated length and the risk of cavitation damage in relation to the characteristics of the gabions.

Al-Fawzy et al. [32, 33] focused on the hydraulic jump formed at the toe of stepped gabion weirs. Their analysis of own experimental data showed the influence of the gravel size, step dimensions, gabion length, and the flow discharge on the distance of the hydraulic jump to the basis of the weir. Salmasi et al. [34] followed previous authors and conducted their study in the sense of better quantifying the energy dissipation in stepped gabion weirs. They tested several geometries involving following variables: step end sill, spillway slope and roughness, and the flow discharge. The roughness was indicated as a relevant parameter in the quantification of the energy dissipation, and the authors presented comparisons of energy dissipations for different roughnesses of the spillway.

Shirvan et al. [35] were able to quantify the effect of a blockage in the upstream surface of a gabion-stepped weir. Flow rate and porosity were used as controlling variables to obtain an empirical equation for the discharge coefficient, presented in nondimensional form.

Singh et al. [36] measured velocity profiles upstream and downstream of gabion-stepped structures proposed to protect coastal regions. The authors considered wave frequencies, geometry of the structures (curved), aggregates' size, which influences the porosity and roughness, and verified the efficiency of the structure for the dissipation of waves energy, mentioning that the obtained results are still in a preliminary stage.

Simões et al. [37] presented an interesting review of the results generated for gabions and proposed a methodology to calculate several aspects of stilling basins. Their study was focused on stilling basins with hydraulic jumps, and the methodology is straightforward to indicate the dimensions and levels of dissipation basins after gabion-stepped weirs.

1.3 Permeable one-step studies: Gabion step

Mohamed [38] studied one-step gabions and presented results for three gravel granulometries, comparing them with impermeable steps. The results involved water depth profiles, dimensionless flow discharges, discharge coefficients, and the conditions of free and submerged flows. The author [38] pointed that the flow occurs over the weir and through the weir, and that for the same discharge the impermeable weir produces a higher water depth than the gabion, which decreases by increasing the gravel size. This implies that equations for impermeable weirs cannot be applied for the free flow over gabions, unless a corrective discharge coefficient is also applied.

Al-Mohammed and Mohammed [5] presented a classification of the flows related to gabions as:

1. Non-overflow (through flow): occurs when the water flows only through the upstream surface of the porous weir.

2. Non-overflow (through flow) limit: is the flow condition in which the upstream free surface coincides with the upper edge of the weir crest.

3. Transition flow: is the condition in which part of the water flows over the porous weir, but is totally absorbed before attaining the end of the weir.

4. Overflow limit: is the flow condition for which the free surface just reaches the downstream edge of the weir crest.

5. Overflow: occurs when there is a stream running over the top of the weir.

The authors studied the effect of the gravel mean size d, the length/height/width of the gabion ($L/H/W$), and the specific flow rate q in the establishment of the upstream depth h_u. Results were presented for the type 1 and for the type 3 flows, obtaining, respectively, Eqs. (4) and (5):

$$h_u = 1.607\frac{q^{0.66}L^{0.349}}{d^{0.34}} \tag{4}$$

$$h_u = 1.624q^{0.196}H^{0.704}L^{0.002} \tag{5}$$

Pallavi and Harshith [39] executed experiments for through flows and overflows, in the sense of obtaining qualitative information about the comparative performance of different experimental conditions. Like Ref. [38], their conclusions testified that the characteristics of impermeable weirs and gabion weirs are different mainly because of the division between overflow and through flow. When applying the same flow rate, the depth attained over the impermeable weir is bigger than that over the gabion weir, the last being smaller for larger gravel diameters. Also, the reduction of velocity is bigger for gabions than for impermeable weirs.

Safarzadeh and Mohajeri [40] considered the traditional equation for discharge through weirs, $q \; \alpha \; h^{3/2}$, analyzing different discharge coefficients proposed in the literature, and finally providing new coefficients valid for specified ranges of nondimensional parameters. Their study was numerical, using experimental data of the literature to calibrate and consubstantiate their simulations.

Focused on the energy dissipation in one-step gabions, Al-Fawzy and Al- Mohammed [41] considered five gabions with different lengths and measured the energy dissipation along each of them. Further, the authors performed a dimensional analysis and adjusted different sets of coefficients for each reach of gravel diameters used in their experiments. Their conclusions consider the interdependence between the length of the gabion, mean dimension of the gravel, the applied unitary discharge, and the observed energy dissipation.

Jalil et al. [42] considered the application of gabions for purposes of partial water storage and spreading of flood waves. They measured the reduction of the flow depths h, the upstream flow depths h_u, the variation of coarseness, discharges Q, gabion heights H, and mean diameters d of the gravel. Equation (6) was presented for the type 1 flow, and Eqs. (7) and (8) for the type 4 flow:

$$h_u = 12.907\frac{Q^{0.618}}{d^{0.328}} \tag{6}$$

$$h_u = 2.441Q^{0.191}H^{0.656}d^{-0.041} \tag{7}$$

$$q = \left[0.015 + 0.193\frac{h}{H} + 0.096\frac{d}{H}\right]\sqrt{g}h_u^{\frac{3}{2}} \tag{8}$$

Shariq et al. [43] developed an equation for flow types 1 and 2 by taking the Ergun's equation and substituting the head loss by the derivative of the water depth in relation to the distance, and by substituting the velocity by the flow rate (discharge) divided by the water depth and the channel width (cross-sectional area). The integration of the obtained differential equation for the water depth produced an implicit equation for the flow rate (discharge), which was solved numerically. The comparison between measured and calculated values produced good results. This way of calculating the discharge through the gravel is understood to be better related to the physical situation of the interstitial flow than the direct use of the free flow weir with corrective coefficients. In this study, this point of view was adopted to obtain the flow rates and water depth evolutions for turbulent and laminar flows separately in the porous matrix.

Salmasi et al. [44] performed own experiments and compared their data and empirical equations with data of the literature. The authors proposed several correlations for the gabion discharge coefficient for free flow and submerged flow conditions, in a procedure that used the free-flow weir equation with corrective coefficients. To obtain the regression equations, multivariate regression analysis and General Practice (GP) methods were used. The authors mentioned reasonable agreement with previous studies found in the literature for both the free flow over impermeable weirs and the free flow over porous weirs.

Applications of gabions together with reviews of the literature were presented by Shah and Gupta [45] and Alsubih et al. [46], while Salehi and Azimi [47] presented an extensive study on how to quantify the discharge through and over gabions using the point of view of the free-flow weir equation with corrective coefficients. The authors proposed friction factors of the Darcy-Weisbach equation for different gravels and the discharge coefficient for different flow conditions. The application of the empirical equations produced results that fit well with observed data.

2. Single gabion rockfill hydraulics

The conceptual study of gabion rockfill hydraulics must follow the flow types mentioned in the Introduction. **Table 1** shows the usual five types of flow, in which the frequently named "transition flow" is renamed to "upper through flow," because it presents the same features of the front through flow, not characterizing a transition

N	Acronym	Name	Description
1	F	Front through	The water flows through part of the upstream face of the porous weir.
2	FL	Front through limit	The water flows through the whole upstream face of the porous weir.
3	U	Upper through	Part of the upper surface is considered.
	PU	Pure upper through	The water flows only through part of the upper face of the porous weir.
	MU	Mixed upper through	The water flows through the upstream face and part of the upper face of the porous weir.
4	UL	Upper through limit	The whole upper surface is considered.
	PUL	Pure upper through limit	The water flows only through the whole upper face of the porous weir.
	MUL	Mixed upper through limit	The water flows through the whole upstream and upper faces of the porous weir.
5	O	Overflow	The water flows over the upper surface of the porous weir like an impermeable weir.

N is the classification number.

Table 1.
Classification of flows for gabions that partially retain water.

between two different kinds of flows. Further, considering experiments in which the upstream face may be impermeabilized, and also for the simplification of derivation purposes, the flow types 3 and 4 of **Table 1** were subdivided into pure and mixed upper flows.

2.1 Mathematical formulation - Basic equations - Physical principles

Following the procedures of Schulz et al. [48] for the hydraulics of cavities in porous media, the conservation principles of mass, momentum, and energy were taken in their integral forms, and focused in steady flows, as shown in Eqs. (9)–(11), respectively:

$$\int_A \rho \vec{V} \bullet d\vec{A} = 0 \tag{9}$$

$$\int_A \vec{V} \rho \vec{V} \bullet d\vec{A} = 0 \tag{10}$$

$$\dot{Q} - \dot{W} = \int_A \left(\frac{V^2}{2} + gz + \frac{p}{\rho} + \mu \right) \rho \vec{V} \bullet d\vec{A} = 0 \tag{11}$$

A is the area crossed by the flow, V is its velocity, ρ the density of the fluid, z is the level, p is the pressure, μ is the internal energy, $\dot{Q}$ is the heat power consumed, and $\dot{W}$ is the work power applied to the control volume.

2.2 Front through and front through limit flows

Both the F and FL flows of **Table 1** use the same kind of control volume, as shown in **Figures 1** and **2**. **Figure 1** shows the difference related to the upstream water level for both flows. This analysis considers the general symbols of **Figure 2**. In this sense,

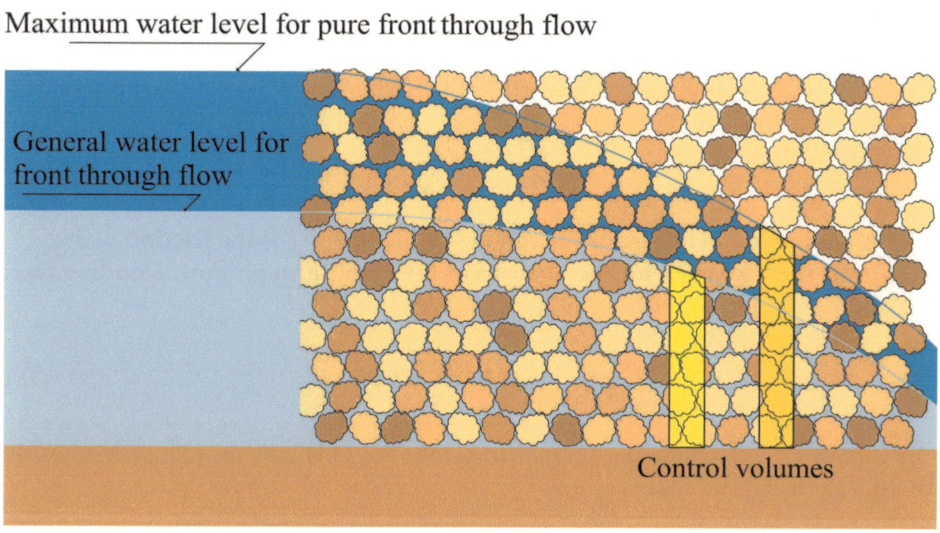

Figure 1.
Flow conditions and control volumes for front through flows.

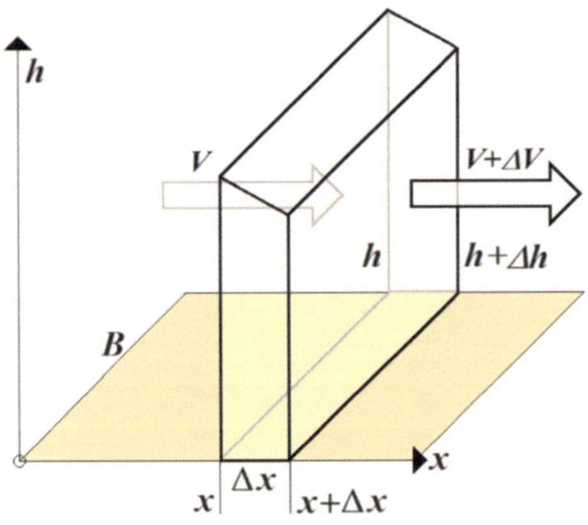

Figure 2.
Detail of the control volume for front through flows.

x, h, and V are, respectively, the position, water depth, and water velocity at the upstream face of the control volume. Further, $x + \Delta x$, $h + \Delta h$, and $V + \Delta V$ are the same variables at the downstream face of the control volume. B is the width of the control volume.

Considering mean values, applying Eqs. (9) and (10) to the control volume of **Figure 2**, considering the hydrostatic pressure distribution as the main cause of motion (not inserting reactive forces in the momentum equation), taken the limits $\Delta x \to 0$ and $\Delta h \to 0$ produces, respectively:

$$\frac{dV}{dh} = -\frac{V}{h} \; or \; h = \frac{q}{V} \tag{12}$$

$$-\rho g h dh = -\rho V^2 dh + 2\rho V h dV \tag{13}$$

q is the specific flow rate (flow rate per unit width). By reducing Eq. (11) to the Bernoulli equation with losses, and using the Darcy-Weisbach equation for the losses, so that resistive components imply in head losses, it results:

$$\frac{dh}{dx} = -f\frac{V^2}{2gD} \tag{14}$$

D is the apparent diameter of the interstitial flow, and f is the friction factor. Eq. (13) was then adjusted for the analysis of the front through flow. Dividing this equation by $\rho\, dx$, we obtain:

$$-gh\frac{dh}{dx} = +V^2\frac{dh}{dx} + 2Vh\frac{dV}{dx} \tag{15}$$

By taking $h = q/V$ from Eqs. (12), dh/dx from Eq. (14) and coupling these results with Eq. (15) it follows that:

$$\frac{dV}{dx} - f\frac{V^4}{4Dgq} - f\frac{V}{4D} = 0 \tag{16}$$

Equation (16) is the governing equation of the front through flow for the present analysis. A normalized form is proposed in Eqs. (17) and (18) for fully turbulent flows, for which f is considered a constant value:

$$\frac{du}{ds} - au^4 - au = 0 \tag{17}$$

$$s = \frac{x}{L} \quad and \quad u = \frac{V}{(gq)^{1/3}} = F_{hor}^{2/3} \quad and \quad a = \frac{fL}{4D} \tag{18}$$

F_{hor} is the horizontal Froude number, and s and u are the normalized distance and velocity, respectively. The fully turbulent solution of Eq. (17) is:

$$u = \frac{1}{\sqrt[3]{K_1 e^{-3as} - 1}} \quad or \quad \frac{u}{u_0} = \frac{\sqrt[3]{K_1 - 1}}{\sqrt[3]{K_1 e^{-3as} - 1}} \tag{19}$$

u_0 represents u at $s = 0$. For turbulent flows, the graph of u as a function of s is a family of nondimensional curves that depends on K_1 (integration constant) and a. Eqs. (19) are similar to the usual equations for free weirs. Coupling Eqs. (12), (18), and (19) it is obtained that:

$$q = \frac{\sqrt{g}\, h^{3/2}}{\sqrt[2]{K_1 e^{-3as} - 1}} \quad or\, q = C^* \sqrt{g}\, h^{3/2}, C^* = \left(K_1 e^{-3as} - 1\right)^{-1/2} \tag{20}$$

The specific flow rate is proportional to the depth elevated to the power 3/2 and the acceleration of gravity elevated to the power 1/2, like in free weirs. The discharge coefficient, C^*, is a function of the position along the length of the gabion. And, because q is constant along the gabion, h is related to s through Eqs. (21):

$$h = \frac{q^{2/3}\sqrt[3]{K_1 e^{-3as} - 1}}{\sqrt[3]{g}} \quad or \quad \frac{h}{h_0} = \frac{\sqrt[3]{K_1 e^{-3as} - 1}}{\sqrt[3]{K_1 - 1}} \tag{21}$$

h_0 is the depth at $s = 0$, while h is the depth at a generic position s. As an additional note, the use of a reactive drag force proportional to V^2 in Eq. (13) produces results similar to the further Eq. (24). This procedure was not detailed here, considering the interesting results already obtained from the analyses of Eqs. (19)–(21).

Figure 3 shows a graph containing experimental data of Refs. [5, 42], and compared with the basic result of Eqs. (20) for $s = 0$, for which $h = h_0$. The figure shows the partial superposition of the regions of data corresponding to $L = 0.23$, 0.30, and 0.60 m. The straight lines in orange, blue, and red are particular empirical linear regressions of Refs. [5, 42] for different gravel diameters d, which are restricted to the sets of data of each regression (not general results). The conclusion that $q\ \alpha\ h_0^{3/2}$, which follows the know relation for free weirs, is evident in the graph. The coefficient of proportionality of course depends on L and d. The authors [5, 42] also presented empirical regression equations for which approximately $q \sim h_0^n$, $n \sim 0.6$ to 0.7, giving support to the theoretical more general and physically supported Eqs. (20).

Equations (21) were used to compare their predictions of h/h_0, with the equation proposed by Stephenson [2], shown in the graph. The same boundary conditions were used for both equations, that is, $h/h_c = h_o/h_c = 2$ for s = 0 and $h/h_c = 1$ for s = 1. The graph of **Figure 4** shows compatible trends for $a \sim 0.01$. Further, Eq. (21) is explicit,

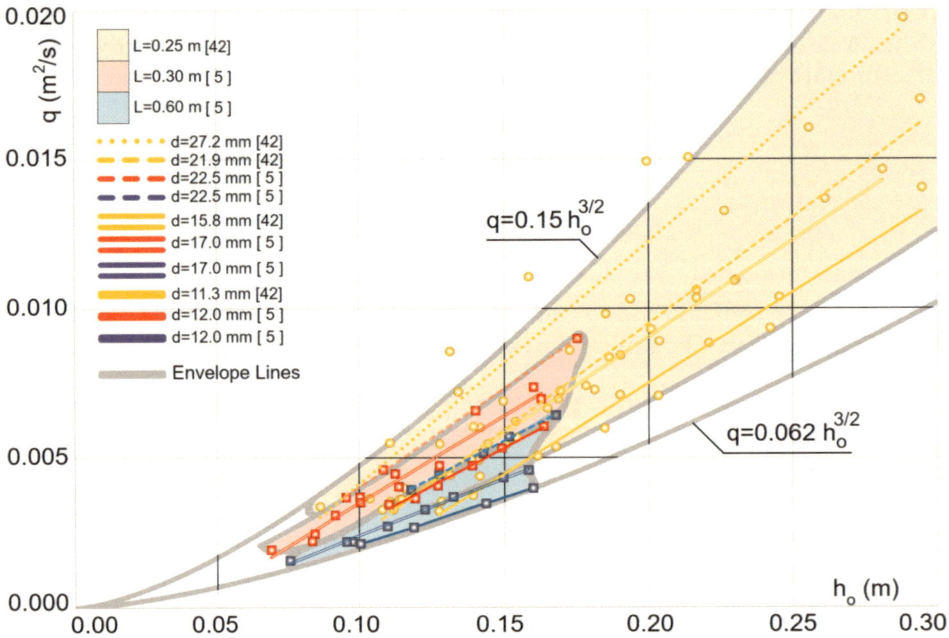

Figure 3.
$q \alpha h_o^{3/2}$ from Eqs. (20) evidenced for front through flows. The data were digitally recovered from Refs. [5, 42]. The shown empirical linear lines were taken from the cited literature and are superfluous by having the theoretical Eqs. (20).

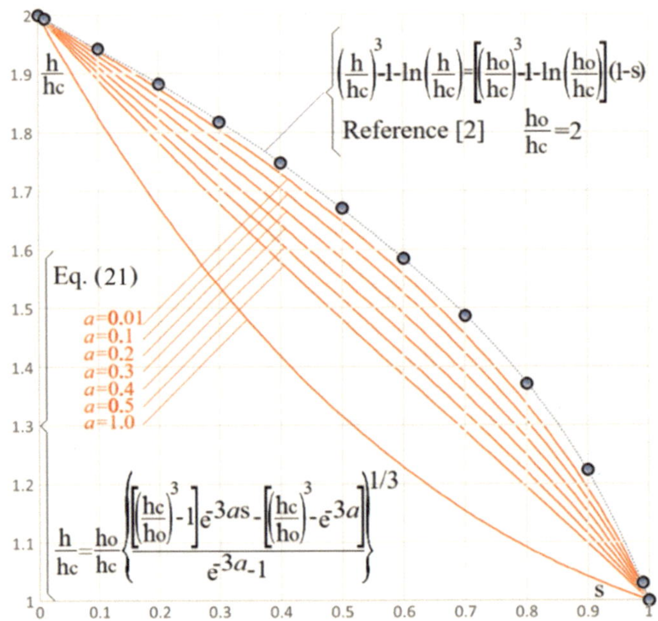

Figure 4.
Comparison of evolution of h/h_o with s from Eq. 21 and Ref. [2].

while Stephenson [2] is an implicit equation, and the present equation accommodates evolutions evolving from convex to concave, depending on the value of a, a characteristic not observed in Stephenson [2].

Taking again Eq. (16), and considering the knowledge of fully laminar flows in tubes, f is a function of the Reynolds number in the form $f = 64/Re_D$, where $Re_D = VD/\nu$ and ν is the kinematic viscosity of the fluid. Considering the same expression of f also valid in Eq. (16) we obtain Eqs. (22) and (23):

$$\frac{du}{ds} - bu^3 - b = 0 \tag{22}$$

$$b = \frac{16}{Re_D}\frac{L}{D} \quad \text{and} \quad Re_D = \frac{(gq)^{1/3}D}{\nu} \tag{23}$$

The solution of Eq. (22) is given by:

$$-ln\left(\frac{u^2 - u + 1}{u^2 + 2u + 1}\right) - \sqrt{3}\ arctan\left(\frac{2u - 1}{\sqrt{3}}\right) = 6bs + K_2 \tag{24}$$

Solving for s we have an explicit relation between s and u, in the form:

$$s = \frac{1}{6b}\,ln\left(\frac{u^2 + 2u + 1}{u^2 - u + 1}\right) - \frac{\sqrt{3}}{6b}\ arctan\left(\frac{2u - 1}{\sqrt{3}}\right) + K_3 \tag{25}$$

The value of u that produces a maximum value of s is $u_M = (4-\sqrt{3})/(4+\sqrt{3}) \approx 0.40$ for any values of b and the integration constant K_3. In this sense, a theoretical control value of the mean laminar velocity in the porous media under the restrictions of the flow in gabions is:

$$V = \left(\frac{4 - \sqrt{3}}{4 + \sqrt{3}}\right)(gq)^{1/3} = 0.8470\ \sqrt[3]{q} \tag{26}$$

Equations (26) and (18) lead to a dependence between the depth and the flow rate similar to free weirs, that is:

$$q = (0.8470)^{3/2}h^{3/2}\ or\ q = 0.7795\,h^{3/2} \tag{27}$$

If it is intended to use Eqs. (26) and (27) as control equations, a good choice is to impose $s = 1$, that is, to locate the maximum distance x at the downstream section of the gabion (which practical use must still be verified). Using $u = (4-\sqrt{3})/(4+\sqrt{3})$ into Eq. (25) leads to Eq. (28) for the maximum s:

$$s \cong \frac{2.140}{b} + K_3 \tag{28}$$

Remembering that b is given by Eqs. (23), for the position $s = 1$, the constants b and K_1 are simply related as:

$$K_3 \cong 1 - \frac{2.140}{b} \tag{29}$$

2.3 Upper through and upper through limit flows

The mass conservation Eq. (9) applied now to the element of volume of **Figures 5** and **6** produces:

$$Vh = (V + \Delta V)(h + \Delta h) + V_y \Delta_x \tag{30}$$

$$V_y = -\left(V \frac{\Delta h}{\Delta_x} + h \frac{\Delta V}{\Delta_x}\right) \tag{31}$$

For $\Delta x \to 0$:

$$V_y = -\frac{d(Vh)}{dx} \tag{32}$$

For the first step of the momentum quantification, Eq. (10) was applied to the same volume of **Figures 5** and **6** in the x (horizontal) direction, and produces:

$$\frac{\rho g h^2}{2} - \frac{\rho g (h + \Delta h)^2}{2} - \tau \Delta_x = -\rho V^2 h + \rho (V + \Delta V)^2 (h + \Delta h) + \rho V V_p \Delta_x \tag{33}$$

$$-\rho g h \frac{\Delta h}{\Delta_x} - \tau = 2\rho h V \frac{\Delta V}{\Delta_x} + \rho V^2 \frac{\Delta h}{\Delta_x} + \rho V V_y \tag{34}$$

For $\Delta x \to 0$:

$$-\rho g h \frac{dh}{dx} - \tau = 2\rho h V \frac{dV}{dx} + \rho V^2 \frac{dh}{dx} + \rho V V_y \tag{35}$$

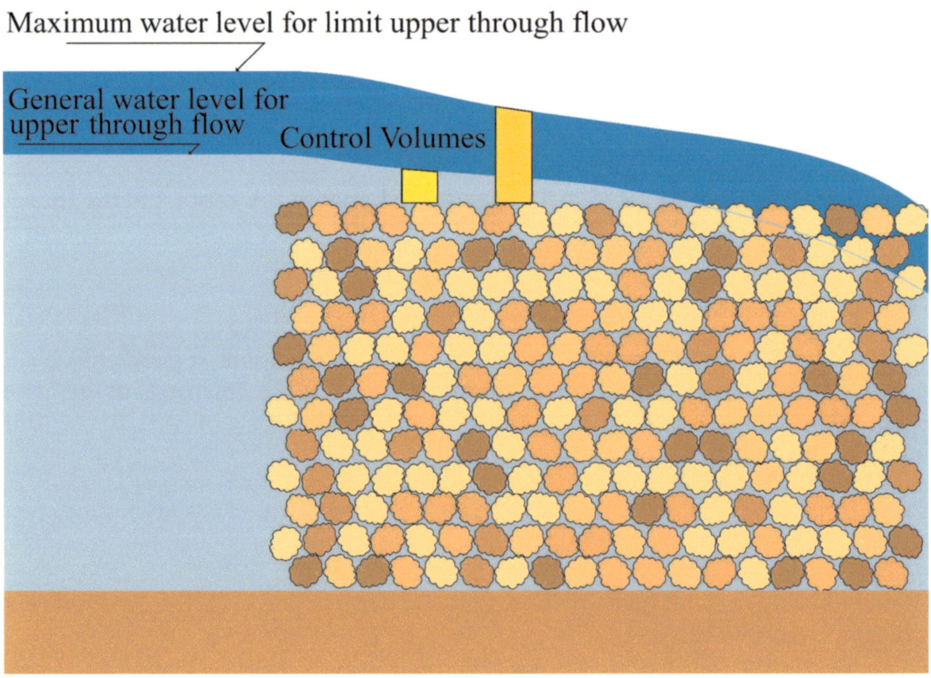

Figure 5.
Flow conditions and control volumes for upper through flows.

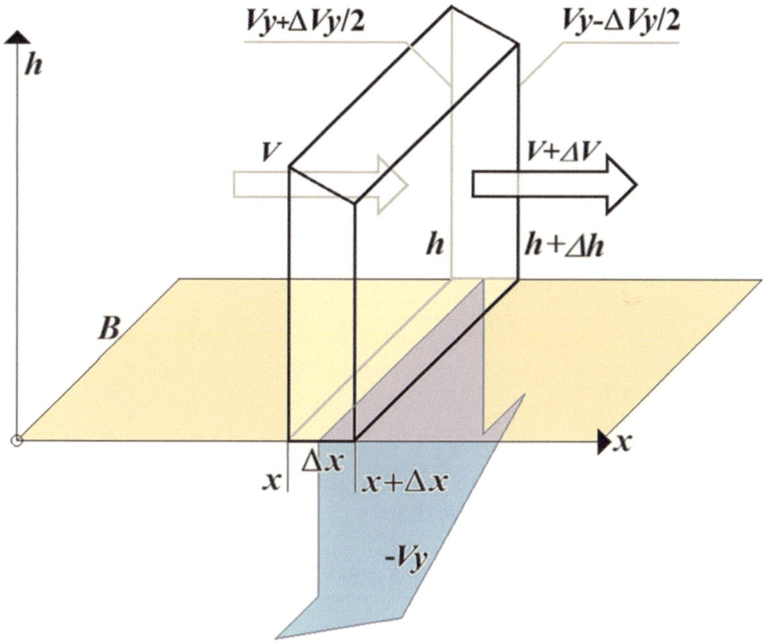

Figure 6.
Detail of the control volume for upper through flows.

Taking the shear stress as a function of the horizontal drag coefficient C_D:

$$-\rho g h \frac{dh}{dx} - \rho C_D V^2 = 2\rho h V \frac{dV}{dx} + \rho V^2 \frac{dh}{dx} + \rho V V_y \tag{36}$$

Coupling Eqs. (32) and (36), and rearranging:

$$-gh \frac{dh}{dx} - C_D V^2 = \frac{h}{2} \frac{dV^2}{dx} \tag{37}$$

Equation (37) is the governing equation of the x direction of the momentum.

For the second step of the momentum quantification, Eq. (10) was applied to the same volume of **Figures 5** and **6** in the y (vertical) direction, and produces, disregarding small second- and third-order parcels:

$$F_y = \rho V\left(V_y + \frac{\Delta V_y}{2}\right)h - \rho(V + \Delta V)\left(V_y - \frac{\Delta V_y}{2}\right)(h + \Delta h) - \rho V_y^2 \Delta_x \tag{38}$$

$$F_y = \rho V h \Delta V_y - \rho V_y h \Delta V - \rho V V_y \Delta h - \rho V_y^2 \Delta_x \tag{39}$$

Taking the force F_y generated by vertical movements in the same way as the shear stress of the horizontal bottom flow movements, we have:

$$F_y = 2\rho h C_V \Delta V_y^2 \tag{40}$$

ΔV_y is the difference between the velocities of the centers of the control volume and the neighbor vertical slices of flow (with similar length Δx). C_V is a vertical drag coefficient. Like before, disregarding small second-order terms, we have:

$$F_y = 0 \tag{41}$$

Coupling Eqs. (39) and (41), then dividing by Δx and applying $\Delta x \to 0$ we have:

$$0 = -\left(Vh\frac{dV_y}{dx} + V_y\frac{dVh}{dx}\right) - V_y^2 \tag{42}$$

Coupling Eqs. (32) and (42) resultas in:

$$Vh\frac{dV_y}{dx} = 0 \tag{43}$$

In this case, the result is:

$$V_y = Constant \tag{44}$$

From Eqs. (32) and (44) it follows that:

$$Vh = K_4 - V_y x \tag{45}$$

K_4 is an integration constant, and the evolution of the specific flow rate $q = Vh$ is linear in x. This result is the basic equation of momentum for the y direction.

The Energy equation was applied in the form of the Bernoulli equation with local and distributed head losses for mean cross-sectional velocities in the gravel (**Figure** 7), leading to:

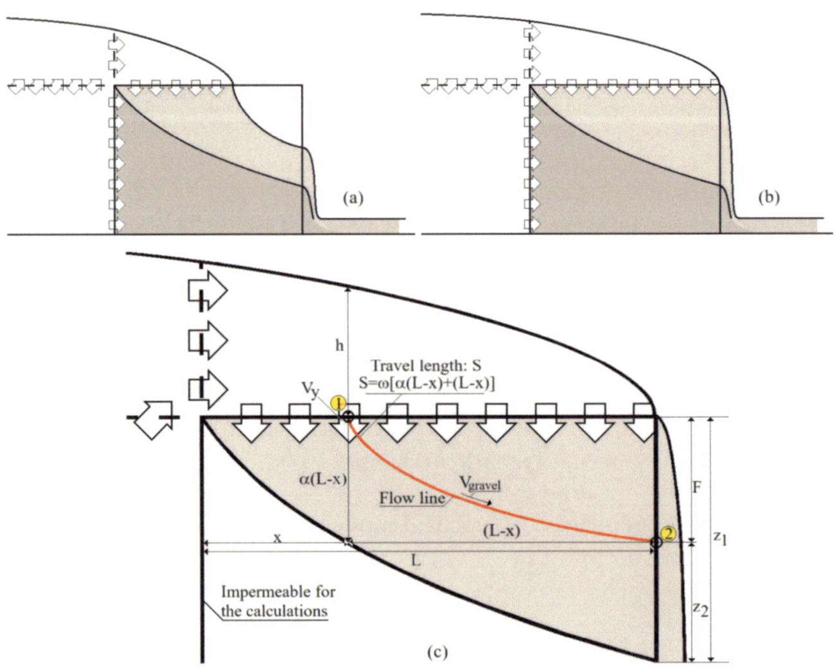

Figure 7.
Sketch of flow through situations in gabions, and definition of variables for Eq. (46). (a) MU flow type; (b) MUL flow type; (c) PUL flow type with the variables of Eq. (46).

$$z_1 + \frac{p_1}{\rho g} + \frac{V_{Gravel}^2}{2g} = z_2 + \frac{p_2}{\rho g} + \frac{V_{Gravel}^2}{2g}(1 + K_{local}) + f\frac{S}{D}\frac{V_{Gravel}^2}{2g} \tag{46}$$

$$h + F = \frac{V_{Gravel}^2}{2g}\left(K_{local} + f\frac{S}{D}\right) \tag{47}$$

K_{local} is a localized head loss coefficient. V_{Gravel} is the characteristic velocity in the gravel (rockfill). Taking the travel length S as a linear function of the vertical ($F = \alpha (L\text{-}x)$) and horizontal ($L\text{-}x$) travel lengths of the water portions at the lower border of the flow, Eq. (48) is obtained:

$$h + \alpha(L - x) = \frac{V_{Gravel}^2}{2g}\left[\frac{f\ \omega[(L - x) + \alpha(L - x)] + K_{local}D}{D}\right] \tag{48}$$

α and ω are constant coefficients. For $x = 0$ the boundary conditions impose that $h = h_0$ and $V_{Gravel} = V_y$. Equation (48) then produces:

$$h_0 + \alpha L = \frac{V_y^2}{2g}\left[\frac{f\ \omega L(1 + \alpha)}{D} + K_{local}\right] \tag{49}$$

$$V_y = \sqrt{\frac{2g(h_0 + \alpha L)D}{f\ \omega L(1 + \alpha) + K_{local}D}} \tag{50}$$

Equation (44) for the vertical component of the momentum points V_y as a constant. Equation (50) shows that this constant depends on active components of the flow and external conditions (h_0, which imposes the initial pressure and g), and the reactive components of the gravel ($D, f, K_{local}, L, \alpha, \omega$).

Further, in this study the final condition of upper through flows is to have $h = 0$ in some position of the upper surface. Taking the value $h = h_L = 0$ at the downstream end of the surface ($x = L$), we have from Eq. (48):

$$h_L = \frac{K_{local}V_L^2}{2g} \tag{51}$$

V_L is the velocity in the gravel at position L. To have $h_L = 0$, it is necessary to have $K_{local} = 0$, so that localized losses are irrelevant in relation to distributed losses. Equations (50) and (51) thus produce:

$$V_y = \sqrt{\frac{2g(h_0 + \alpha L)D}{f\ \omega L(1 + \alpha)}} \tag{52}$$

This result shows that the boundary conditions define the values of the parameters used in the formulation (like V_y) and relations between integration constants and parameters defined for the calculations. Different boundary conditions can be studied and are very welcomed.

Equation (37) for the horizontal component of the momentum is now retaken to verify possible evolutions of the surface level of the surface of the water. In this study, it was considered that C_D approaches the laminar condition for slower flows, being inversely proportional to the Reynolds number $Re_L = VL/\nu$. This produces, with some rearrangements

$$-gh\frac{dh}{dx} - \frac{C_L\nu}{L}V = hV\frac{dV}{dx} \tag{53}$$

hV is known from Eq. (45), which, when applied to Eq. (53), gives us:

$$-gh\frac{dh}{dx} - \frac{C_L\nu}{L}V = \left(K_4 - V_y x\right)\frac{dV}{dx} \tag{54}$$

Defining $s = K_4 - V_y x$, taking Eq. (45) as $V = s/h$ and rearranging produces:

$$V_y gh\frac{dh}{ds} - \frac{C_L\nu}{L}\frac{s}{h} = -\frac{V_y s}{h} + V_y\left(\frac{s}{h}\right)^2\frac{dh}{ds} \tag{55}$$

Multiplying Eq. (55) by ds/dh, taking now $s = Vh$ and rearranging leads to:

$$\frac{dV^2}{dh} + \frac{2C_L\nu}{\left(C_L\nu - LV_y\right)}\frac{V^2}{h} = \frac{2LgV_y}{\left(C_L\nu - LV_y\right)} \tag{56}$$

Defining the constant coefficients (V_y is constant from Eq. (44)):

$$A = \frac{2C_L\nu}{\left(C_L\nu - LV_y\right)} \quad B = \frac{2LgV_y}{\left(C_L\nu - LV_y\right)} \tag{57}$$

Equations (56) and (57) show that the differential Eq. (56) is linear for V^2, being then easily integrated, whose result is shown in Eq. (58).

$$V = \sqrt{\frac{B}{A+1}h + \frac{K_5}{h^A}} \tag{58}$$

K_5 is an integration constant. Because the evolution of Vh is known from Eq. (45), also the evolution of h is known when coupling Eqs. (45) and (58):

$$K_4 - V_y x = h\sqrt{\frac{B}{A+1}h + \frac{K_5}{h^A}} \tag{59}$$

The boundary condition $h = h_0$ at $x = 0$, and $h = 0$ at $x = L$ (for this study) produce $K_5 = 0$ and $K_4 = V_y L$, so that Eq. (59) converts to:

$$h = \frac{\left[V_y(L-x)\right]^{2/3}}{\left(\frac{B}{A+1}\right)^{1/3}} \tag{60}$$

Other boundary conditions can be tested and are very welcomed for the study of rockfill hydraulics. In normalized form Eq. (60) is written as shown in Eqs. (61) and (62).

$$h^* = (1-s)^{2/3} \tag{61}$$

$$h^* = h\left[\frac{B}{(A+1)\left(V_y L\right)^2}\right]^{-1/3} \tag{62}$$

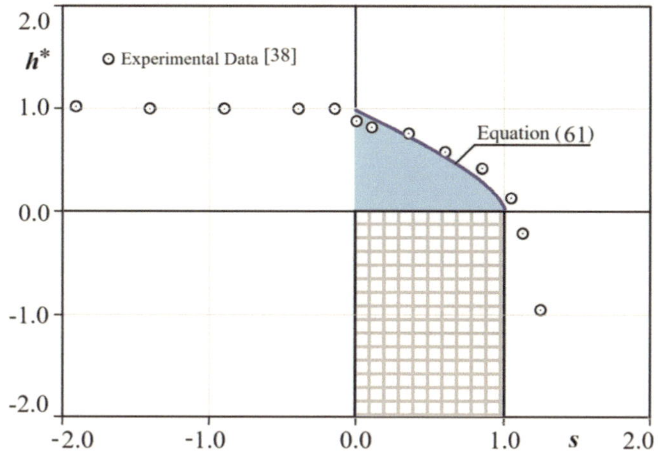

Figure 8.
Equation (61) compared with data of Ref. [38] digitally recovered.

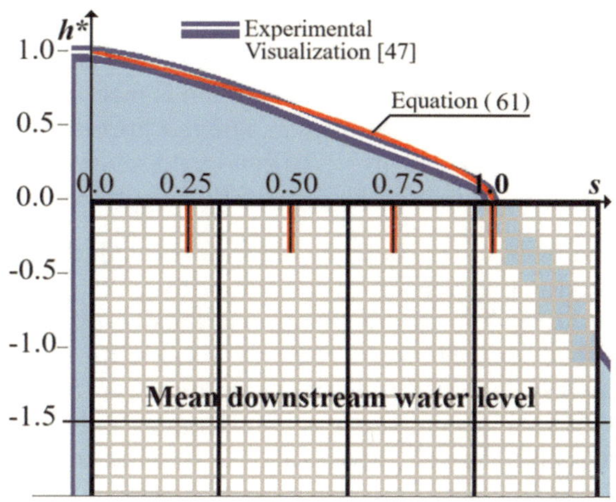

Figure 9.
Equation (61) compared with a digitally recovered profile of a photograph visualization of Ref. [47].

Equation (61) is an elegant 2/3 power law of the deficit of the distance (1-s). It was compared with the data of the literature for situations of upper through flow (or very close to it). The comparisons are shown in **Figure 8** (approximation to PUL) and **9** (approximation to PU), and the theoretical profile is adequate to be used as reference for the measured profiles. Considering the experimental data disposable in the literature, it was observed that more experimental studies are needed to better represent the different flow conditions of the classification of **Table 1**, so that the theoretical equations can be used as reference with more security.

3. Challenges

Having followed classical derivation procedures and obtained a set of equations for several aspects of through flow in single gabion steps, some questions may be pointed for additional studies.

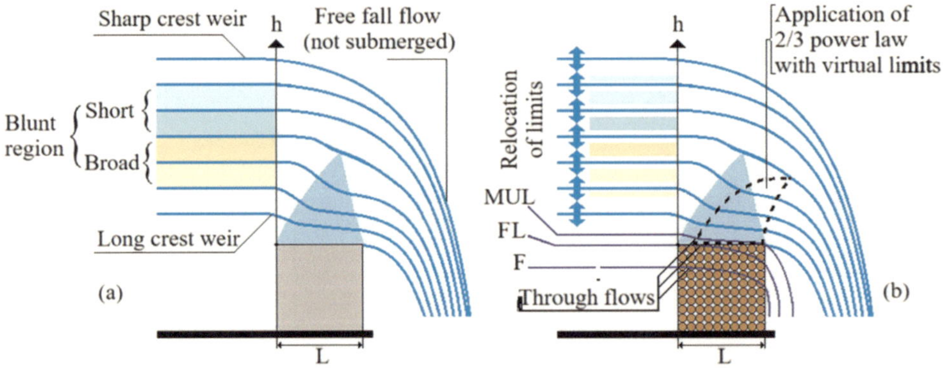

Figure 10.
Schematic and qualitative representation of the 4 crest types for (a) impermeable weirs [14], and (b) the need of a redefinition for permeable weirs.

- Solutions for front through and pure upper through flows were presented. It remains to provide the solutions for mixed upper through flows, named "transition flows" in some of the former literature. These mixed flows are a sum of the separate contributions of front through and pure upper through flows. The new aspect to be considered for the combined solution is that the front through flow is now subjected to a higher impulsive force (a higher pressure acting on the front area, due to the higher water depth). The combined solution will allow quicker checks with data of measured discharges and flow depths of the literature.

- In this study, the internal resistance to the flow through the gravel was considered by taking the inclination of the water surface (or energy line) provided by the Darcy-Weissbach equation in the one-dimensional energy equation. A further possible insertion of resistance in the set of equations, also explained in the text, is to add a resistive drag force (with respective drag coefficient) in the momentum equation. For laminar and turbulent flows, the drag is proportional to the velocity elevated to the first and second powers, respectively. The adequacy of this insertion of resistance must be tested by comparing predictions and measurements. This providence will add security in the application of the physical concepts to rockfill hydraulics.

- Azimi and Rajaratnam [14] studied impermeable weirs with finite crests and proposed a four types classification considering the length of the crest: sharp; short; broad and long-crested weirs. Among various parameters considered by Azimi and Rajaratnam [14], the lateral aspect ratio h/L showed to be of importance, being h the head on the weir, and L its length. **Figure 10a** and **b** shows impermeable and permeable weirs, respectively, and the limiting trends to sharp and to long crests in **Figure 10a**. Considering the modified dependences between discharges and water depths, it is also of interest to verify how to adjust the classification of Azimi and Rajaratnam [14] to permeable weirs (**Figure 10b**).

4. Conclusions

Through flows for one-step gabion weirs were theoretically analyzed. The physical concepts of mass, momentum, and energy conservation were used. The now obtained equations for flow depths and discharges reproduce empirical data and qualitative observations of the literature, showing that they are more general than the former empirical regressions. Further, comparing flow depths of a theoretical literature proposal and this study shows that the new equation produces convex and concave depth profiles, as well as adequately follows the literature profile for proper values of its control parameter, and is an explicit equation (the literature equation is implicit). Thus, the new conceptual equation embraces a broader range of trends.

A further contribution was made for the classification of the through flows, allowing the proper application of the physical principles of conservation (mass, moment, energy) to obtain the prediction equations. The upper through flows were subclassified into pure upper through flows and mixed upper through flows.

For the front through flows, the obtained equation for the specific discharge has a form similar to the free weir equation, but with the discharge coefficient expressed as a function of the position. The dependence between the discharge and the water depth as a 3/2 power law is theoretically maintained, even for the flow occurring through the gravel. Comparisons with empirical data of the literature substantiated this dependence. Also, theoretical predictions of the evolution of the velocity and the water depth with the distance were obtained, and justify previous literature data.

For the pure upper flows, predictions for velocities, specific discharges, and water depths were also obtained. Comparing literature data and visualizations for the evolution of the water depth, the here obtained 2/3 power of the deficit of the length showed to be a good reference equation. It was pointed that more experimental studies will be welcomed to compare predictions and measurements. The good results suggest the continuity of theoretical studies based on physical principles. The section "Challenges" points to several most immediate extensions, which will be welcomed by the engineers and researchers of this field of Engineering Sciences.

Data availabilty statement

All data, models, or code that support the findings are available from the corresponding author by request.

Notation

The following symbols are used in this paper.

A	area
a	normalized coefficient ($a = fL/4D$)
B	width
b	normalized coefficient (b = 16 $L/Re_D D$)
C^*	coefficient of flow rate equation
C_D	drag coefficient
C_V	vertical drag coefficient
d	gravel diameter
D	apparent diameter for flow in gravel
F_1	supercritical Froude number
F_y	vertical force
f	friction factor
g	acceleration of gravity
h	water depth

h_u	upstream water depth
h^*	normalized water depth
H	depth (height) of gabion
L	length of gabion
$K_{1\ to\ 5}$	integration constants
p	pressure
Q	flow rate
$\dot{Q}$	heat crossing a control surface
Re_D	Reynolds number ($Re_D = VD/\nu$)
S	travel length
s	normalized distance ($s = x/L$)
u	normalized velocity
V	velocity
V_y	vertical velocity
W	gabion width
$\dot{W}$	work crossing a control surface
x, y	horizontal and vertical axes
$y_{1,2}$	sequent depths 1 and 2
z	vertical axis
Δh	h interval
ΔV	V interval
Δx	distance interval taken along x
$\phi_{1,2}$	coefficients of sequent depths
ρ	water density
μ	internal energy
τ	shear stress

Author details

Harry Edmar Schulz[1*] and André Luiz Simões[2]

1 Auburn University, Auburn, USA

2 Federal University of Bahia, Salvador, Brazil

*Address all correspondence to: hes0071@auburn.edu

References

[1] Charman J, Kostov L, Minetti L, Stoutesdijk J, Tricoli D, Berney O. Small Dams and Weirs in Earth and Gabion Materials. Rome: Food and Agriculture Organization of the United Nations, Land and Water Development Division; 2001. 179 p. (AGL/MISC/32/2001)

[2] Stephenson D. Rockfill in Hydraulic Engineering. Developments in Geotechnical Engineering. Amsterdam: Elsevier Scientific Publishing Company; 1979. 230 p

[3] Neermal M. The use of gabions in hydraulic applications. In: Civil Engineering. 2012;4 p. Available from: http://www.saice.org.za/downloads/monthly_publications/2012/2012-Civil-Engineering-June/#/0 [Accessed: October 1, 2025]

[4] Thompson DM, Puklin LS, Marshall AE. The long-term impact of channel stabilization using gabion structures on Zealand River, New Hampshire. Ecological Engineering. 2016;**95**:779-792. DOI: 10.1016/j.ecoleng.2016.07.016

[5] Al-Mohammed FM, Mohammed SH. Flow through and over Gravel Gabion Weirs. Al-Furat Al-Awsat Technical University and University of Mustansiriya; 2015. 13 p. Available from: https://www.researchgate.net/publication/323401667_FLOW_THROUGH_AND_OVER_GRAVEL_GABION_WEIRS [Accessed: October 1, 2025]

[6] Gill MA. Hydraulics of rectangular vertical drop structures. Journal of Hydraulic Research. 1979;**17**(4):289-302. DOI: 10.1080/00221687909499573

[7] Schulz HE, Nóbrega JD, Simões ALA, Schulz H, Porto RM. Details of hydraulic jumps for design criteria of hydraulic structures. In: Schulz HE, editor. Hydrodynamics—Concepts and Experiments. London: IntechOpen; 2015. DOI: 10.5772/58963

[8] Schulz HE, Simões ALA, Nóbrega JD. Roller lengths, sequent depths, surface profiles for pre-design of dissipation basins. In: Proceedings of the 2nd International Workshop on Hydraulic Structures: Data Validation. Vol. 1. Coimbra, Portugal: IAHR; 2015. pp. 81-93. DOI: 10.13140/RG.2.1.3027.0889

[9] Schulz HE, Vasconcelos JG, Patrick AC. Air entrainment in pipe-filling bores and pressurization interfaces. ASCE Journal of Hydraulic Engineering. 2020;**146**(2):04019053. DOI: 10.1061/(ASCE)HY.1943-7900.0001672

[10] Rajaratnam N, Chamani MR. Energy loss at drops. Journal of Hydraulic Research. 1995;**33**(3):373-384. DOI: 10.1080/00221689509498577

[11] Wu S, Rajaratnam N. Impinging jet and surface flow regimes at drop. Journal of Hydraulic Research. 1998;**36**(1):69-74. DOI: 10.1080/00221689809498378

[12] Chamani MR, Beirami MK. Flow characteristics at drops. ASCE Journal of Hydraulic Engineering. 2002;**128**(8):788-791. DOI: 10.1061/(ASCE)0733-9429(2002)128:8(788)

[13] Chamani MR, Rajaratnam N, Beirami MK. Turbulent jet energy dissipation at vertical drops. ASCE Journal of Hydraulic Engineering. 2008;**134**(10):1532-1535. DOI: 10.1061/(ASCE)0733-9429(2008)134:10(1532)

[14] Azimi AH, Rajaratnam N. Discharge characteristics of weirs of finite crest

length. ASCE Journal of Hydraulic Engineering. 2009;**135**:1081-1085. DOI: 10.1061/(ASCE) HY.1943-7900.0000117

[15] Amin MR, Rajaratnam N, Zhu DZ. Flow and energy loss downstream of rectangular sharp-crested weirs for free and submerged flows. Journal of Fluids Engineering, Transactions ASME. 2022; **144**:031201-6. DOI: 10.1115/1.4052049

[16] Dust D, Wohl E. Characterization of the hydraulics at natural step crests in step-pool streams via weir flow concepts. Water Resources Research. 2012;**48**:W09542. DOI: 10.1029/ 2011WR011724

[17] Kozeny J. Über kapillare Leitung des Wassers im Boden (Aufstieg, Versickerung und Anwendung auf die Bewässerung). Sitzungsberichte der Akademie der Wissenschaften, mathematisch-naturwissenschaftliche Klasse. 1927;**136**(2a):271-306

[18] Carman PC. Fluid flow through granular beds. Transactions of the Institution of Chemical Engineers. 1927; **12**:150-166

[19] Carman PC. Flow of Gases through Porous Media. London: Butterworths Scientific Publications, Academic Press; 1956. 182 p

[20] Bugeat L. Les ouvrages en gabions: techniques rurales en Afrique. Secretariat d'Etat aux Affaires Etrangeres: France; 1969. 113 p. Available from: https://www.ircwash.org/sites/default/files/201-2702.pdf [Accessed: October 1, 2025]

[21] Peyras L, Royet P, Degoutte G. Ecoulement et dissipation sur les déversoirs en gradins de gabions. La Houille Blanche. 1991;**1**:37-47. DOI: 10.1051/lhb/1991003

[22] Peyras L, Royet P, Degoutte G. Flow and energy dissipation over stepped gabion weirs. ASCE Journal of Hydraulic Engineering. 1992;**118**(5):707-717. DOI: 10.1061/(ASCE)0733-9429(1992) 118:5(707)

[23] Brunet G, Shuey R. Stream Bank Stabilization with Vegetated Gabions. Maccaferri: Inc./Gove Environmental Services, Inc.; 2006. Available from: https://www.academia.edu/1527172/STREAM_BANK_STABILIZATION_WITH_VEGETATED_GABIONS [Accessed: October 1, 2025]

[24] Chinnarasri C, Donjadee S, Israngkura U. Hydraulic characteristics of gabion-stepped weirs. ASCE Journal of Hydraulic Engineering. 2008;**134**(8): 1147-1152. DOI: 10.1061/(ASCE) 0733-9429(2008)134:8(1147)

[25] Khatibi R, Salmasi F, Ghorbani MA, Asadi H. Modelling energy dissipation over stepped-gabion weirs by artificial intelligence. Water Resources Management. 2014;**28**:1807-1821. DOI: 10.1007/s11269-014-0545-y

[26] Rand W. Flow geometry at straight drop spillway. Proceedings of the American Society of Civil Engineers. 1955;**81**(Paper 791):1-13. DOI: 10.1061/ JYCEAJ.0000031

[27] Wüthrich D, Chanson H. Aeration and Energy Dissipation over Stepped Gabion Spillways: A Physical Study. Hydraulic Model Report No. CH92/13. Brisbane: School of Civil Engineering, University of Queensland; 2014. 171 p

[28] Wüthrich D, Chanson H. Air entrainment and energy dissipation on gabion stepped weirs. In: Proceedings of the 5th International Symposium on Hydraulic Structures and Society: Engineering Challenges and Extremes.

Brisbane, Australia: IAHR; 2014. 9 p. DOI: 10.14264/uql.2014b.12

[29] Minho TRP, Simões ALA, Queiroz LM, Arcanjo GS, Schulz HE, Porto RM, et al. Numerical modeling and empirical formulation for re-aeration in nappe flow without hydraulic jump in stepped chutes. RBRH. 2024;**29**:1-10. DOI: 10.1590/2318-0331.292420240070

[30] Velázquez-Luna L, Ventura-Ramos E. Concentrated flow through gabions weirs. In: Conference of the Universidad Autónoma de Querétaro. 2017. 978-1-5386-1694-9/17/IEEE. Available from: https://ieeexplore.ieee.org/document/7968188

[31] Reeve DE, Zuhaira AA, Karunarathna H. Computational investigation of hydraulic performance variation with geometry in gabion stepped spillways. Water Science and Engineering. 2019;**12**(1):62-72. Available from: http://creativecommons.org/licenses/by-nc-nd/4.0/

[32] Al-Fawzy AM, Al-Merib FH, Al-Mohammed FM, Hummadi AH, Al-Zubaidy RZ. Effect of using stepped gabions on the distance of the hydraulic jump. IOP Conference Series: Materials Science and Engineering. 2020;**888**:012061. DOI: 10.1088/1757-899X/888/1/012061

[33] Al-Fawzy AM, Al-Taee KN, Al-Mohammed FM, Hommadi AH. The hydraulic jump formed downstream a stepped gabion weir: An experimental study. In: 3rd International Conference on Recent Innovations in Engineering. IOP Conference Series: Materials Science and Engineering. Vol. 978. Duhok, Iraq: IOPScience; 2020. p. 012018. DOI: 10.1088/1757-899X/978/1/012018

[34] Salmasi F, Razi S, Abraham J. Hydraulics of flow in gabion stepped spillways: An experimental study. Journal of Hydraulic Structures. 2023;**9**(2):75-100. DOI: 10.22055/jhs.2023.40421.1257

[35] Shirvan SH, Pirzadeh B, Rajaei SH, Bejestan MS. Physical hydraulic modeling of gabion stepped weirs with upstream blockage. Iranian Hydraulic Association Journal of Hydraulics. 2023;**18**(2):87-99. DOI: 10.30482/jhyd.2023.369582.1624

[36] Singh AK, Dave M, Salvi R, Juneja A. An experimental investigation into the use of three stepped gabion walls for coastal protection works using acoustic Doppler velocimeter. International Journal of Geosynthetics and Ground. 2024;**10**(1):1-9. DOI: 10.1007/s40891-023-00510-6

[37] Simões ALA, França TIS, Queiroz LM, Lima-Neto IE, Schulz HE, Fontes DCH, et al. Dimensionless equations for hydraulic jump stilling basins downstream of gabion stepped chutes. Journal of Hydraulic Structures. 2024;**11**(1):41-54. DOI: 10.22055/jhs.2024.47500.1313

[38] Mohamed HI. Flow over gabion weirs. Journal of Irrigation and Drainage Engineering. 2010;**136**(8):573-577. DOI: 10.1061/(ASCE)IR.1943-4774.0000215

[39] Pallavi M, Harshith L. Open channel flow characteristics using gabion weir. International Research Journal of Engineering and Technology. 2018;**5**(12):928-933

[40] Safarzadeh A, Mohajeri SH. Hydrodynamics of rectangular broad-crested porous weirs. ASCE Journal of Irrigation and Drainage Engineering.

2018;**144**(10):04018028. DOI: 10.1061/(ASCE)IR.1943-4774.0001338

[41] Al-Fawzy AM, Al-Mohammed FM. Dissipation of energy of flow by conventional type of gabion weir. IOP Conference Series: Materials Science and Engineering. 2019;**584**:012038. DOI: 10.1088/1757-899X/584/1/012038

[42] Jalil SA, Sarhan SA, Hussein BS, Qasim JM. Effect of gravel size and weir height on flow properties of gabions. Journal of University of Babylon for Engineering Sciences. 2019;**27**(2): 214-222. DOI: 10.29196/jubes.v27i2.2340

[43] Shariq A, Hussain A, Ahmad Z. Discharge equation for the gabion weir under through flow condition. Flow Measurement and Instrumentation. 2020;**74**:101769. DOI: 10.1016/j.flowmeasinst.2020.101769

[44] Salmasi F, Sabahi N, Abraham J. Discharge coefficients for rectangular broad-crested gabion weirs: Experimental study. ASCE Journal of Irrigation and Drainage Engineering. 2021;**147**(3):04021001. DOI: 10.1061/(ASCE)IR.1943-4774.0001535

[45] Shah MC, Gupta KK. Review on flow characteristics of gabion weir. In: Proceedings of the International Conference on Advancements in Engineering and Sciences (ICAES). 2021. AIP Conference Proceedings. Dehradun, India: AIP Publishing; 2481. p. 020043. DOI: 10.1063/5.0105385

[46] Alsubih M, Ahmed M, Alqadhi S, Mallick J. Gabion water barrier structures as a sustainable approach to water and land conservation. Environmental Science and Pollution Research. 2023;**30**:126057-126071. DOI: 10.1007/s11356-023-31040-3

[47] Salehi S, Azimi AH. Hydraulic characterization of turbulent flow through and over a porous media. Flow Measurement and Instrumentation. 2024;**96**:102533. DOI: 10.1016/j.flowmeasinst.2024.102533

[48] Schulz HE, van Zyl JE, Yu T, Lima-Neto IE, Souza-Filho FA, Correa NA, et al. Hydraulics of fluidized cavities in porous matrices: Cavity heights and stability for upward water jets. ASCE Journal of Hydraulic Engineering. 2021;**147**(10):04021037. DOI: 10.1061/(ASCE)HY.1943-7900.0001911

Section 3

Human Flow Scales and Natural Flow Scales

Chapter 4

Supercritical Flow in Junctions of Drainage Galleries/Channels: Case Study – Granjeiro River, City of Crato, Brazil

Francisco Osny Enéas da Silva

Abstract

In urban drainage systems, it is common for storm drains to discharge into macro-drainage channels with the junction occurring at strong angles, close to 90°. These storms drain junctions in channels may be uneven and differ hydraulically from the concept of supercritical junctions between leveled channels and at weak angles below 40°. This chapter suggests the concept of hydraulic puncture for junctions between galleries and channels at strong angles, both of which may be in a supercritical regime or with the main channel in a supercritical regime and the gallery in a forced conduit regime. The submerged or semi-submerged jet coming from the gallery would be the hydraulic puncture. Theoretical considerations are made based on a case study. A simplified, purely vectorial methodology is suggested as a first approximation for design purposes, and three-dimensional modeling using computational fluid dynamics (CFD) is considered indispensable to obtain greater precision without having to resort to experimental tests.

Keywords: drainage channels, junction between storm drains and channels, hydraulic puncture, junction flow, choking flow, supercritical channels

1. Introduction

The drainage of rainwater in urban areas requires a system of micro-drainage works consisting of gutters, manholes, connection lines, and storm drains that collect surface runoff water, leading it to macro-drainage axes, which are formed by thalwegs that can be rivers, streams, or artificial channels, lined or not. The focus of this chapter is on the combination of storm drains with lined urban channels.

The discharge of water from storm drains into lined urban channels normally occurs at high angles, close to 90°, due to the road layout of the streets that reach the macro-drainage axes in which the drains are built, in addition to being a junction at different levels, normally with the bottom of the drain located from a few centimeters to even meters above the bottom of the macro-drainage channel.

IntechOpen

These conditions, considered normal for junctions of galleries in urban drainage channels, have been little studied, with almost no references in the specialized literature. Research has focused on junctions of secondary or lateral channels in main or straight-flow channels, at angles equal to or less than 60°, leveled on a horizontal bottom, and with a rectangular cross-section [1]. The flow regimes studied ranged from subcritical, critical, transitional, and supercritical regimes. Research has shown that the junction of channels in a supercritical regime is a hydraulic phenomenon of an essentially three-dimensional (3-D) nature; however, efforts have been made by researchers to simplify the process of hydraulic calculations from empirical formulations for 1-D and/or 2-D, as can be seen in Bellamoli et al. [2].

In this chapter, we will specifically address the problem of storm drains in lined urban canals, both flowing in a supercritical regime. The flow from the gallery entering an urban canal, both at high speed, in a situation of drowning by the main canal, generates such a powerful shock jet that we suggest adopting the term hydraulic puncture.

The force of the supercritical jet in the hydraulic puncture, even when submerged, can destroy the main channel in addition to the junction itself.

This phenomenon occurred simultaneously in six storm drain junctions with the urban channel lined with the Granjeiro River that crosses the city of Crato, in the state of Ceará, Brazil, on January 28, 2011, during an intense rainfall of 162 mm occurring in just four hours. The first breach verified in the drainage system is the case study addressed here.

2. Current knowledge about channel junctions

The merging of shallow water channels to form flow networks in channels is a common occurrence in urban rainwater drainage systems or in irrigation projects. The confluence of channels has been the subject of research since the 1940s, although there are few studies in the specialized literature on the subject.

The first study we were able to identify on the rise in water levels at the confluence of river channels dates to 1944 [3], in which expressions were developed to calculate the height of water levels at the junction of channels in a subcritical regime of the same width and horizontal bottom, with junction angles of 45° and 135°, with good experimental results being obtained only for the 45° angle. The equations were derived from the principle of conservation of linear momentum applied to a control volume surrounding the junction, for which simplifying hypotheses were assumed, such as the hydrostatic distribution of pressures, the neglect of friction forces, and the consideration of equal water levels and pressures in the upstream sections of both channels, the tributary side, and the main channel in a straight passage. The model is still used today in the design of junctions.

The junction of channels in which at least one of them flows in a supercritical regime, that is, with a Froude number greater than unity ($Fr > 1$), acquires characteristics that are different from those of junctions in which both channels flow in a subcritical regime.

According to Chow [4], for the case of channels with a high slope, the supercritical flow of uniform depth will break into a train of moving waves or pulses known as pulsating flow. Any disturbance in the supercritical flow gives rise to standing waves associated with the concentration of flow [1, 4].

Supercritical flows at junctions form standing waves of great height due to the collision of two torrential flows from upstream [1]. Due to the essentially three-dimensional nature of the collision between the two supercritical flows at the junction, there is still no precise and generalized analytical model for solving the problem of channel junctions in the supercritical regime due to the large number of parameters that interfere in the determination of the waterline profile, velocities, and pressures on the channel structures.

Research on channel junctions in the supercritical regime has been successively performed using experimental models, such as in Schwalt and Hager [5]; Greated [6]; Christodoulou [7]; Hager [8, 9]; Schwalt [10]; Weber [11]; and Herty and Seaïd [12], citing here the most relevant ones found in this research.

The most recent advances in supercritical flow in channel junction analysis have occurred from the use of experimental models to evaluate the results of simulations using computational fluid dynamics (CFD).

In Blagojevic et al. [13], simulations were conducted in a T-shaped open channel junction using two different turbulence models: k-ω shear stress transport (SST) and k-ω SST scale-adaptive simulation (SAS), in conjunction with Volume of Fluid (VOF) and mixture multiphase models. The results of time-averaged free surface data obtained from CFD simulation were compared with measurements using Light Detection and Ranging (LIDAR). The results indicated good agreement between the predictions by the CFD model and the measurements with LIDAR for the elevations of the free surface at the junction and its level fluctuation. It was also observed that the k-ω SST model performed very well in the simulated conditions and that there would be no advantage in using the SAS model, which requires greater computational effort.

In Haidar et al. [14], the free surface supercritical flow of channel junctions at right angles of equal width was simulated using the 2-D shallow water equations, employing two calculation methods: in the first approach, termed "uniform," a uniform mesh was used, both in the channels and in the junction. In the second approach, two mesh spacings are adopted: one for the channels and the other for the junction and the adjoining channel lengths, termed "sparse." It was observed that the sparse mesh model can reproduce the typical flow structure in the supercritical channel junction, despite diverging in the recirculation zones. However, the sparse mesh model can be 61% computationally more efficient than the uniform mesh model.

Among the experimental studies, the work of Schwalt and Hager [5] deserves to be highlighted, as they presented a robust conceptualization and formulation for the problem of joining two horizontal rectangular channels, with flow in the lateral channel at an angle δ. **Figure 1** shows the conceptual scheme developed in the research.

According to Schwalt and Hager [5], the entry of a side channel at angle δ into a straight-through channel or main channel, both in the supercritical regime, generates standing waves as shown in **Figure 1**. Wave C is a junction wave located at the rear of the junction point and can be compared with the wave due to an abrupt deflection of the wall. Waves B and D are wall waves located downstream of wave C due to its impact on the wall opposite to the side branch and due to the reflection of wave B on the opposite wall, respectively.

The authors tested experiments with junction angles of 30° and 60°, with channel widths of 300–500 mm in a model large enough to avoid scale effects.

As a result of the experiments, 16 equations were established to calculate the lengths, heights, and start and end points of each type of wave and the geometry of the separation zone.

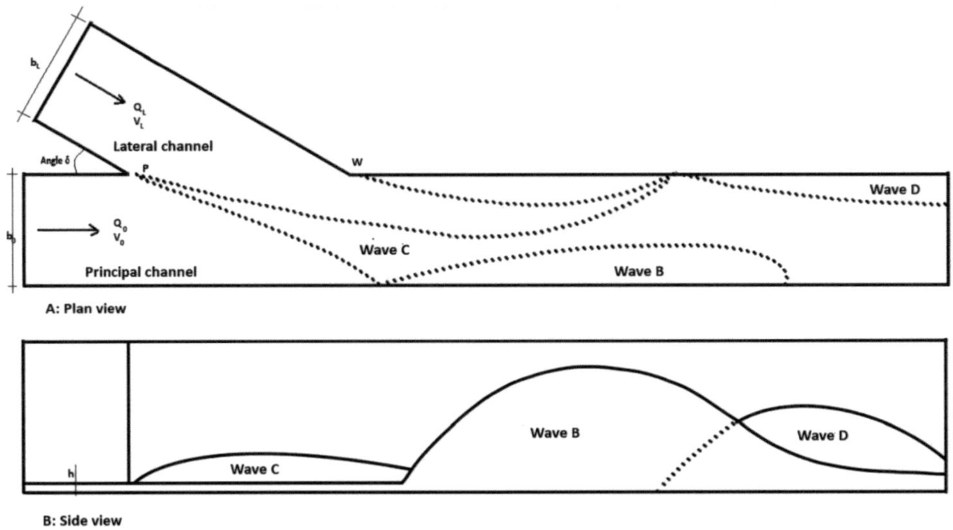

Figure 1.
Supercritical junction flow (schematic). Adapted from Schwalt and Hager [5].

The concept of the determining Froude number given by eq. (1) was presented.

$$f = 2 \cdot \frac{F_0 \cdot F_L}{(F_0 + F_L)} \quad (1)$$

Where F_0 is the Froude number of the flow in the main channel and F_L is the Froude number of the flow in the side channel.

The concept of shock number S given by eq. (2) was also addressed.

$$S = f \cdot sin\delta \quad (2)$$

The shock number S strongly interferes with the formation of the C-type wave. The B-type wave is always the tallest and is independent of the angle δ, while the C and D waves depend on the junction angle δ.

According to Schwalt and Hager [5], the geometry of the junction flow changes significantly for two transition angles:

- (1) transition between parallel branch junction (δ = 0°) and weak junction angle, at $\delta_1 \approx 15°$, and

- (2) transition between weak to strong junction angles, at $\delta_2 \approx 40°$.

In transition (1), the separation zone becomes important, and wall waves can occur.

At transition (2), the impact wave starts to split, and backflow occurs to the upstream branch. Thus, the flow at the junction is strongly disturbed. Further experiments at the 90° junction revealed that a local hydraulic jump was always present in the junction domain, and supercritical junction flow in the present context never occurred.

The junction angle should be limited to $\delta \leq \delta_2$, therefore, supercritical flow is prone to strong losses and pulsations otherwise. Furthermore, a transition to choked flow occurs, and the supercritical flow through the junction breaks down.

3. Case study: Granjeiro River channel in the City of Crato, Brazil

The city of Crato is located in the northeast of Brazil, in the state of Ceará, and has a population of 131,050 inhabitants. The city is crossed by the Granjeiro stream, which originates in a mountain range near the city and runs along a natural bed until the beginning of the urban section of the city of Crato, where the Granjeiro stream becomes channeled in a trapezoidal section lined with concrete.

The channeled urban section of the Granjeiro stream receives inflows from stormwater galleries that flow into the channel at angles ranging from 70 to 90°. The drainage basin of the Granjeiro stream in the outflow section that corresponds to the location of the inlet of the first stormwater gallery has an area of 10.69 km^2. The cross-section of the urban channel of the Granjeiro stream has a base of 3 m, a maximum height of 4 m, an H:V slope of 0.875:1, and a concrete lining. The slope is 0.02877 m/m in the first section of the Granjeiro River channel, which results in a supercritical pulsating flow regime (see **Figure 2**).

Figure 2 shows the onset of runoff due to 64 mm of rainfall on October 19, 2011.

3.1 Case description

During the early morning of January 28, 2011, 162 mm of rainfall occurred in a period of just 4 hours in the drainage basin of the Granjeiro stream in the city of Crato. The channel overflowed, flooding 3200 homes and commercial establishments and causing great damage to the population. The event that attracted the most attention was the rupture of the channel at the junction with all the storm drains that flowed into the channel. The first tributary to the channel that suffered structural collapse is shown in the sequence of **Figures 3–5**.

Figure 2.
Supercritical flow in the Granjeiro stream channel in the initial urban section upstream of the entrance to the first storm drain (The Author, October 19, 2011).

Figure 3.
First collapsed storm drain tributary to the Granjeiro stream, the object of this study (The Author, January 30, 2011).

Figure 4.
Detailed view of the rupture immediately downstream of the entrance to the first storm drain (The Author, January 30, 2011).

Figure 3 shows the first stormwater gallery tributary to the Granjeiro stream channel on the second day after the storm.

Figure 4 shows a closer view of this section of the gallery collapse, looking from upstream to downstream. Note the semicircular pattern of the rupture that affected

Figure 5.
Panoramic view from upstream to downstream of the collapse site of the first collapsed gallery tributary to the Granjeiro stream channel (The Author, January 30, 2011).

both the left bank of the channel (opposite bank to the gallery) and the right bank, immediately downstream of the entrance of the storm drain into the Granjeiro stream channel.

Figure 5 shows a more panoramic view from upstream to downstream of the location of the first rupture in the storm drain inflow, where it can be observed that the exact location of the rupture (indicated by the arrows) was located approximately in the type B wave zone, indicated in **Figure 1**. There was also a rupture in the separation zone that corresponds approximately downstream of the point W shown in **Figure 1**.

The tributary storm drain is 1074.0 m long and drains a basin of 0.849 km^2. The drain is rectangular with the following dimensions: width b = 2.40 m; height h = 1.20 m; longitudinal slope S0 = 0.02873 m/m, practically the same as that of the Granjeiro stream channel. The drain enters the channel at an angle $\delta = 77.83°$.

The bottom of the gallery is 1.5 m above the bottom of the Granjeiro stream channel.

4. Methodology

The study methodology initially included a regressive hydrological modeling of the flows produced by the 162 mm rainfall that occurred on January 28, 2011, to determine the resulting flow, both in the Granjeiro stream channel and in the tributary stormwater gallery, using the Hydrologic Engineering Center-Hydrologic Modeling System (HEC-HMS) software using the Soil Conservation Service (SCS), actual Natural Resources Conservation Service unit hydrograph model.

Next, an analysis of the hydraulic characteristics of the flow was carried out in terms of depth, velocity, and flow regime. To assess the impact of the meeting of the waters at the junction on the banks of the channel, in the absence of a robust and precise theoretical model, a global analysis of the forces produced in a vectorial manner was carried out. Although it does not offer calculation precision, it can serve as an estimate of the forces acting on the opposite bank for short-term design purposes.

Finally, CFD modeling was used to assess whether it would be possible to reproduce at least qualitatively the Schwalt and Hager [5] model for comparison purposes with the case studied.

4.1 Hydrological modeling

The rainfall equation for the city of Crato [15] is as in eq. (3):

$$i = \frac{20{,}240 \cdot (T_r - 2{,}98)^{0{,}0769}}{(t + 5{,}45)^{0{,}723}} \tag{3}$$

Where i is the intensity of the rain in mm/min, t is the duration of the rain in minutes, and Tr is the return time in years.

A hyetogram was composed using the alternating block method, which is shown in **Figure 6**.

The curve number used according to the soil characteristics (type B) and the use and occupation of the basin's soil was CN = 80, AMC-II condition, since the rainfall under analysis occurred within these conditions. The percentage of impermeable area considered was 38% based on previous studies by the author.

The final combined hydrograph that was generated by the HEC-HMS program [16] for the section of interest at the junction of the channel and the gallery is shown in **Figure 7**.

According to the data from the regression analysis performed with HEC-HMS, the peak flow rate passing through the section of interest would have been 334.8 m^3/s, occurring at 05:25 am. The peak flow rate in the Granjeiro stream channel would have been 316.7 m^3/s, also occurring at 05:25 am, while the peak flow rate in the storm drain would have been 39.3 m^3/s, occurring at 05:05 am. Therefore, there may have been a time lag between the peak flows of around 20 minutes. The period between 05:00 am and 05:15 am would have been approximately the time when the here-defined hydraulic puncture occurred in the channel.

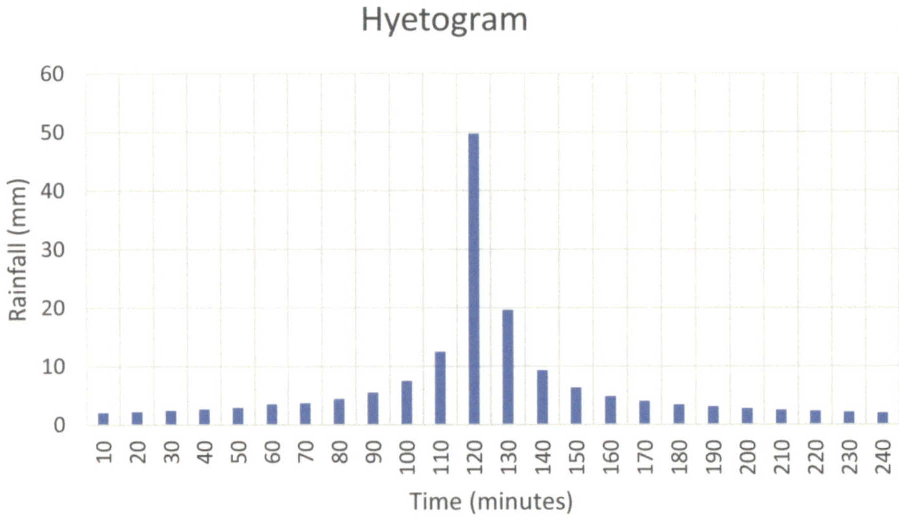

Figure 6.
Hyetogram of the 162 mm rainfall that occurred on January 28, 2011.

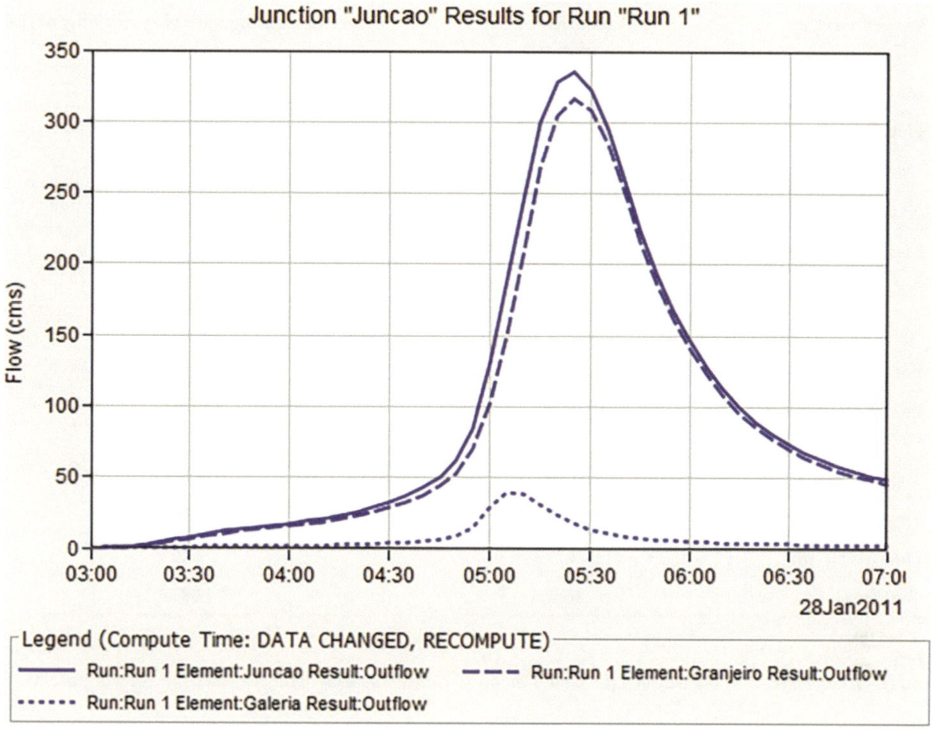

Figure 7.
Combined hydrograph at the junction of the Granjeiro stream channel with the storm drain.

At the time of 05:05 am, which corresponds to the peak flow in the gallery, the flow in the channel would have been 150.2 m^3/s.

4.2 Hydraulic modeling

For the main channel (Granjeiro stream), **Table 1** presents the hydraulic characteristics for both the peak flow rate that occurred at 05:25 am and for the time of 05:05 am, coinciding with the peak flow rate in the gallery.

For the storm drain tributary to the main channel of the Granjeiro stream, the flow promotes flow in a forced conduit regime. **Table 2** presents the data for the peak flow that possibly occurred at 05:05 am on January 28, 2011.

4.2.1 Analysis

The gallery begins to flow in the channel regime, and as the flow increases, it begins to flow as a forced conduit. The transition flow rate from the channel regime to the forced conduit regime in the gallery would have been 24.804 m^3/s and should have occurred at about 04:58 am. At this transition limit, the flow rate of 24.804 m^3/s would correspond to a Froude number of 2.51, confirming the flow was in a supercritical regime. For the forced conduit regime, it makes no sense to use the Froude number parameter, since the Reynolds Number would be more applicable, characterizing a high-speed turbulent regime.

The flow of the gallery at the time of peak flow, in a forced conduit regime, entering the channel at a high angle (δ = 77.83°) and with a step of 1.5 m above the bottom of the

Parameter/Unit	Peak flow for the channel	Flow coinciding with the peak flow of the gallery
Input data		
Flow (m^3/s)	316.7	150.2
Manning coefficient	0.014	0.014
Slope (m/m)	0.02877	0.02877
Bottom width	3.0	3.0
Side slope (H:V)	0.875	0.875
Hydraulic calculations		
Wet area (m^2)	18.891	10.831
Longitudinal slope (m/m)	0.02877	0.02877
Top width (m)	8.667	6.849
Froude number	3.625	3.521
Flow depth (m)	3.238	2.20
Velocity (m/s)	16.765	13.867

Table 1.
Hydraulic characteristics of the channel for peak flows and flow coinciding with the peak flow of the gallery.

Parameter/Unit	Peak flow for the gallery
Input data	
Flow (m^3/s)	39.3
Manning coefficient	0.014
Slope (m/m)	0.02873
Bottom width	2.4
Height-maximum depth	1.2
Side slope (H:V)	
Hydraulic calculations	
Wet area (m^2)	2.88
Longitudinal slope (m/m)	0.02873
Hydraulic diameter (m)	1.60
Reynolds number	2.1832 x 107
Flow depth (m)	1.2
Velocity (m/s)	13.645

Table 2.
Hydraulic characteristics of the gallery for peak flows.

main channel, completely mischaracterizes the case studied in relation to a junction of open channels, which has been researched by the authors cited in the literature review.

The speed of the water leaving the gallery (13.65 m/s) is of the same order of magnitude as the average speed of the channel (13.87 m/s) for the time of peak flow in the gallery, entering the channel in the form of a semi-submerged jet, since the flow depth in the channel is 2.20 m and the step is 1.5 m; thus, the channel has a depth of 0.70 above the bottom of the gallery, which is not enough to leave it completely submerged during peak flow in the gallery.

However, at the time of peak flow in the main channel, the gallery would have been submerged because the depth in the channel became 3.24 m, submerging the gallery 0.54 m above the upper face of the rectangular gallery.

The here-defined hydraulic puncture corresponds to this high-speed submerged or semi-submerged jet from the gallery penetrating at a high angle into the main channel.

4.2.2 Estimation of the jet force by a global vector model

Due to the difficulty of developing a theoretical analytical model that is sufficiently robust and representative of the case described here, considering the three-dimensional behavior of the so-called hydraulic puncture, we propose here for engineers and designers of storm drain junctions in channels an evaluation of the resulting force that the jet promotes on the opposite bank of the channel based on purely vectorial considerations.

The simplified vectorial method consists of determining the centroid of both flows in the channel and in the gallery and performing a vectorial analysis of the vector of the resulting force, using the equation for the force of a fluid on an area normal to the flow.

To use the vector method, the x-axis was aligned with the flow direction of the main channel; the y-axis was aligned with the vector perpendicular to the main channel banks; and the z-axis was the vertical axis pointing upwards. The origin of the coordinate axes was at the intersection between the lines passing through the axis of symmetry of the main channel and the axis of symmetry of the gallery.

The forces Fc and Fg (corresponding to the flows in the main channel and the gallery, respectively) can be determined as a function of the cross-sectional area of the liquid and the flow velocity in each of the structures. Here, the influences of viscous forces and those due to hydrostatic pressure were neglected because they are much smaller in magnitude than the dynamic forces produced by the velocities. Eqs. (4) and (5) define the forces.

$$F_C = \rho\, A_{m,c}\, V_c^2 \tag{4}$$

$$F_G = \rho\, A_{m,g}\, V_g^2 \tag{5}$$

Where Fc and Fg were defined previously, ρ is the specific mass of water, Am is the respective wetted area, and Vc and Vg are the velocities in the main channel and gallery, respectively.

To analyze the displacement and rotation of the fluid at the point of collision between these two forces, we calculated the position vector ($\vec{r}$) and the corresponding scalar distance (|r|) between Fg when leaving the gallery and flowing to the point of impact with Fc (taken as the origin of a local vector field).

For the peak flow rate of the gallery and the corresponding flow rate in the main channel, the centroids of the flows were y_c = 1.24 m and y_g = 2.10 m. The force of the water in the channel was 2082.73 kN, and the force of the water in the gallery was 536.21 kN.

The displacement distances of the gallery force can be determined through a brief geometric analysis in the horizontal and vertical planes. In x and y, the distances were approximately $\Delta x = 0.646$ m and $\Delta y = 3.0575$ m. The elevation on the z-axis is the difference between the centroids y_c and y_g; therefore, $\Delta z = 0.86$ m, such that z is zero at the bottom of the channel. This difference in level produced a rotation to the resulting force of these two flows.

The displacement vector and its magnitude are given by eqs. (6) and (7).

$$\vec{r} = \Delta x\boldsymbol{i} + \Delta y\boldsymbol{j} + \Delta z\kappa \tag{6}$$

$$|r| = \sqrt{\Delta x^2 + \Delta y^2 + \Delta z^2} \tag{7}$$

Thus, multiplying the scalars Fc and Fg by the unit vector $\frac{\vec{r}}{|r|}$, we obtained the vector forces $\vec{F_c}$ and $\vec{F_g}$.

$$\vec{F_c} = 415.13\boldsymbol{i} + 1964.80\boldsymbol{j} + 552.65\kappa \tag{8}$$

$$\vec{F_g} = 106.87\boldsymbol{i} + 505.82\boldsymbol{j} + 142.27\kappa \tag{9}$$

Adding the vector forces $\vec{F_c}$ and $\vec{F_g}$ and calculating their resultant $\vec{F_r}$ we obtain:

$$\vec{F_r} = 522.01\boldsymbol{i} + 2470.62\boldsymbol{j} + 694.92\kappa \tag{10}$$

The resultant module was 2619.95 kN. This force magnitude was enough to cause the collapse of the Granjeiro stream channel structure.

Thus, the resulting vector points to the left bank of the farm stream channel, with the component in the y direction (in the local convention, perpendicular and positive toward the north wall of the channel) being an order of magnitude greater than the others, proving that the jet leaving the gallery promotes this hydraulic puncture on the channel flow, even in a supercritical regime.

The direction of the resultant calculated through this simple vector modeling absolutely agrees with the position of the channel rupture a little further downstream from the gallery entry point, approximately in a position that would correspond to the formation of standing wave B shown in **Figure 1**.

In any case, the premises adopted in Schwalt and Hager [5] differ significantly from the geometric and physical reality of the case studied, not allowing a direct comparison between the studies based on the application of the equations proposed by Schwalt and Hager [5].

4.3 Use of computational fluid dynamics

At the current stage of the art, the most accurate way to analyze hydraulic puncture and determine the resulting forces on the bank of the channel opposite the entrance to the storm drain, which is the part that suffers the greatest impact due to the hydraulic shock between the two flow regimes, would be through experimental hydraulic testing. However, the cost and time spent on experimental testing make their use prohibitive for routine urban drainage projects, where the designer has to make quick design decisions due to the short deadlines for project execution required by most municipalities.

The alternative to experimental testing is the use of computational fluid dynamics (CFD) to perform the analyses. The use of CFD becomes almost mandatory in cases such as the one presented here.

However, CFD modeling is not yet widely accessible to most designers due to the high cost of the software used. There are programs with freely accessible source code, but they are difficult to manipulate, requiring user experience for their correct use.

To exemplify the use of CFD in the modeling of hydraulic puncture in the present case, two software programs (Flow3D Hydro® and SimFlow®) were used to model the flow for the condition of peak flow of the gallery occurring at 05:05 am on January 28, 2011, with the purpose of comparing at least qualitatively with the theory based on the experimental model developed by Schwalt and Hager [5], with regard to the formation of waves immediately downstream of the junction (see **Figure 1**).

For the CFD modeling to be closer to the basic premises of the experimental work of Schwalt and Hager [5], the main channel of the Granjeiro stream, which has a trapezoidal cross-section, was approximated to an equivalent rectangular cross-section, and the bottom was horizontalized. However, the rectangular gallery was maintained with the original rectangular dimensions (2.4 m x 1.2 m), and the bottom was raised 1.5 m in relation to the bottom of the main channel.

Figure 8 shows a cut of the modeling at the moment of maximum flow in the gallery (Qpeak = 39.3 m^3/s) and the main channel with a flow of 150.2 m^3/s for the water layer in the main channel, seeing on the opposite bank the formation of the standing wave B, mentioned in **Figure 1**.

In **Figure 8**, the monitoring slice was placed on the left bank of the main channel precisely with the intention of capturing the type B wave that forms on the bank opposite the junction. It can be seen from the figure that the CFD modeling was able to capture the pulsating flow regime that occurs in the main channel of the Granjeiro stream (see **Figure 2**) and the formation of standing waves within the channel, although the physical arrangement of the junction at the entrance to the gallery in the city of Crato is different from that indicated in the premises of leveling and horizontality at the junction, proposed in the experimental studies by Schwalt and Hager [5].

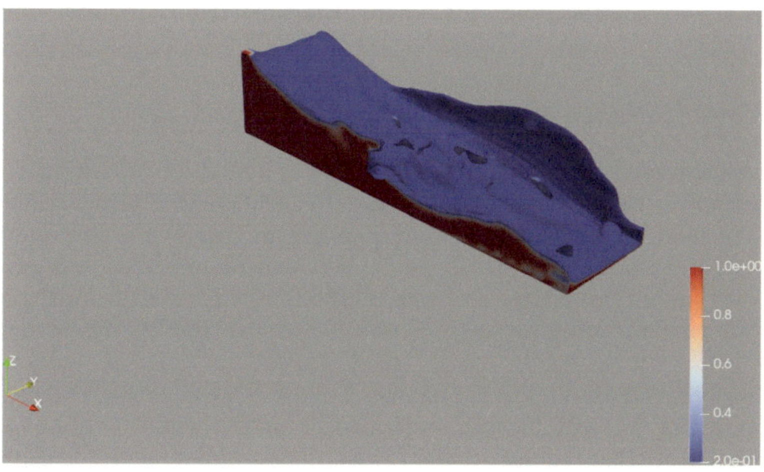

Figure 8.
Instantaneous cut of the moment of discharge of the peak flow of the storm drain into the main channel of the Granjeiro stream, in modeling using CFD.

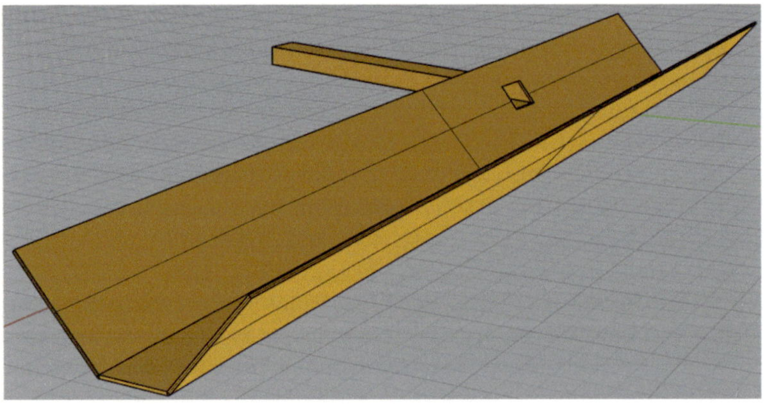

Figure 9.
Perspective view of the real structure model of the gallery/channel junction in the city of Crato, Brazil.

The simulated model presented in **Figure 8** does not correspond to the real structure of the case addressed here for the city of Crato. **Figure 9** presents a perspective view of the real structure of the gallery/channel junction discussed here, whose modeling using CFD has not yet been completed due to the necessity in establishing the appropriate boundary conditions for the passage of the gallery flow from an open channel supercritical regime to the conduit flow regime, drowned by the main channel, causing the phenomenon of hydraulic puncture.

The results obtained so far can qualitatively reproduce the phenomenon; however, the numerical results differ significantly from those observed in the field that promoted the collapse of the gallery/channel junction structure.

Current research into the application of CFD to model the hydraulic puncture concept in the structure shown in **Figure 9** is still ongoing, and the results will be published when completed.

However, although the numerical results obtained using CFD do not yet represent the values observed in the field with reasonable accuracy, this does not invalidate the theme of the chapter, whose main objective is to present the concept of hydraulic puncture in gallery/channel junctions flowing in a supercritical regime.

5. Concluding remarks

This chapter presents the suggestion of adopting a new concept in supercritical channel hydraulics, called hydraulic puncture, to designate the penetration of a submerged or semi-submerged jet at high speed, originating from the junction of a storm drain with a channel at a strong angle, between 70° and 120°, more commonly close to 90°. A simplified, purely vectorial methodology for design approximations was proposed for practical purposes while a more elaborate theory is not yet available.

A case study for an extreme event recorded in the city of Crato-CE, Brazil, served to demonstrate the concept. Recommendations are made for the use of CFD-based modeling for studies that require greater calculation accuracy as a replacement for experimental models, since computational fluid dynamics modeling can reasonably capture the behavior of stationary waves that form due to the collision between two high-speed flows.

The author continues research on the concept of hydraulic puncture using computational fluid dynamics (CFD) modeling to try to arrive at a simplified two-dimensional mathematical formulation that is feasible for rapid use in preliminary design evaluations.

Acknowledgements

The author is indebted to Michal Sajkowski from SimFlow CFD and Luciano Santana from M&N Simulações e Projetos, representatives of Flow-3D in Brazil, for the kind transfer of the provisional license of the SimFlow and Flow-3D software, respectively, to carry out this research.

Author details

Francisco Osny Enéas da Silva
Department of Hydraulic and Environmental Engineering, Federal University of Ceará, Brazil

*Address all correspondence to: osnysilva@ufc.br

References

[1] Coelho MMLP, Souza PA, Baptista MB. Investigações experimentais em confluência de canais. Curitiba, Brazil: XV Simpósio Brasileiro de Recursos Hídricos; 2003. ISSN 2318-0358

[2] Bellamoli F, Müller LO, Toro EF. A numerical method for junctions in networks of shallow-water channels. Applied Mathematics and Computation. 2018;**337**:190-213

[3] Taylor EH. Flow characteristics at rectangular open-channel junctions. Transactions ASCE. 1944;**109**:893-912

[4] Chow VT. Open channel hydraulics. New York: McGraw Hill; 1959

[5] Schwalt M, Hager WH. Experiments to supercritical junction flow. Experiments in Fluids. 1995;**18**:429-437

[6] Greated CA. Supercritical flow through a junction. La Houille Branche. 1968;**8**:693-695

[7] Christodoulou GC. Incipient hydraulic jump at channel junctions. Journal of Hydraulic Engineering. 1993;**119**(3):409-421

[8] Hager WH. Supercritical flow in channel junctions. Journal of Hydraulic Engineering. 1989a;**115**:595-616

[9] Hager WH. Transitional flow in channel junctions. Journal of Hydraulic Engineering. 1989b;**115**:243-259

[10] Schwalt M. Vereinigung schiessender abflüsse (Supercritical flow at junctions). PhD thesis 10370. Zurich: Swiss Federal Institute of Technology ETH; 1994

[11] Weber LJ, Schumate ED, Mawer N. Experiments on flow at a 90° open-channel junctions. Journal of Hydraulic Engineering. 2001;**127**(5):340-350

[12] Herty M, Seaïd M. Assessment of coupling conditions in water way intersections. International Journal for Numerical Methods in Fluids. 2013;**71**:1438-1460

[13] Blagojevic M, Hocevar M, Bizjan B, Dresar P, Repinc SK, Rak G. Three-dimensional numerical simulation of a two-phase supercritical open channel junction flow. Water. 2024;**16**:1757. DOI: 103390/w16121757

[14] Haidar S, Gabriel HF, Khan SA. Supercritical flow simulation at a right channel junction. Comparison between a uniform and a sparse mesh. Journal of Civil Engineering. 2017;**21**(7):2984-2990. DOI: 10.1007/s12205-017-0811-7

[15] Fechine V. Aplicação do método das isozonas na obtenção das equações IDF de chuvas intensas dos municípios de Juazeiro do Norte, Barbalha e Crato – CE. Fortaleza-CE: Universidade Federal do Ceará; 2011. 72 p

[16] United States Army Corps of Engineers. Hydrologic Modeling System HEC-HMS. 4.11. 2003. User's Manual

Chapter 5

Measuring and Modeling Sediment Transport: Case Study of a Brazilian Cerrado (Savanna) River

Carlos Eugenio Pereira

Abstract

The damming of water for the formation of multipurpose reservoirs causes changes in sediment transport and deposition processes, potentially leading to various operational and environmental impacts. The Jordão River watershed, the study area of this research, has the potential for electricity generation to supply the Triângulo Mineiro and Alto Paranaíba regions in Brazil. Understanding sediment transport processes is essential, as it significantly contributes to ensuring proper water resource management in the region. This research aims to investigate the sediment transport dynamics in this watershed, seeking to provide a preliminary understanding of the behavior of solid particles within this water body. This study involved five sampling stations, with eight measurement campaigns conducted to assess liquid discharge and collect water for suspended solids and bed sediment analysis during the dry and rainy seasons, covering the period from September 2018 to May 2019. Several analyses were carried out, including: the characterization of bed sediment granulometry; analysis of the hydraulic and geometric characteristics of the studied stations; application of indirect methods for calculating solid discharge; use of the Halley-Smith equipment; and application of the HEC-RAS program for sediment transport, through which it was possible to understand the dynamics of sediment transport during the analyzed period. Contrary to expectations, no significant increase in sediment transport was observed with rising river discharge. Granulometric curve analyses of bed sediments revealed the deposition of finer sediments during the dry season, especially at the stations located in the Jordão River.

Keywords: sedimentology, sediment load, liquid discharge, Halley-smith, HEC-RAS

1. Introduction

In electric power generation, one of the first hydroelectric investments made in Brazil was the implementation of Small Hydroelectric Power Plants (SHPs), which are still widely used today. The adoption of these small plants is explained by their efficiency and speed of energy production (decentralized production with minimal losses in transmission systems); in addition to requiring low initial investment, they have a lower environmental impact, while needing shorter periods for obtaining licenses, as well as for civil construction [1].

Considering that 60% of the annual electricity production in Brazil comes from hydroelectric sources [2], where reservoirs are subject to siltation with sediments generally originating from the upstream watershed, which intensifies abrasion through their suspended particles in turbines and machinery, means that research related to this topic becomes relevant, especially since electric energy is essential for the development of the country.

It is known that among the main factors related to the increase in environmental degradation in watersheds, which intensifies negative environmental impacts and can cause extensive and even irreversible damage, directly affecting natural balance, are the frequent unplanned changes in land use, with the indiscriminate removal of vegetation from the upstream watershed of hydroelectric plants. This practice leaves the soil unprotected, and erosive processes increase as surface runoff intensifies. Thus, the lack of watershed planning and the inadequate use of natural resources lead to water contamination and the intensification of erosive processes [3], resulting in soil erosion and increasing the sediment load in the river. After erosion occurs, these sediments reach the water body, where they are transported in suspension or along the riverbed until they are deposited.

Erosion, transport and sediment deposition are natural environmental phenomena; however, these have been significantly intensified and accelerated by human activity [4].

The sediment load in rivers is an indicator of the erosive processes occurring within the watershed [5]. In addition to factors such as precipitation, Mohammad et al. [6] and Silva et al. [7] demonstrate that the sediment load in rivers is influenced by soil type, vegetation cover, land use and watershed topography. In the river, due to the high energy of the current flow, sediments are transported, but when this current reaches a particular reservoir, the flow slows down and the sediment is deposited.

In sedimentological studies, direct methods are more complex and costly; therefore, the use of indirect methodologies is more common. The most recognized methods can be found in Beling et al. [8]. These indirect methods were applied in the studies of Sobrinho et al. [9] and Brune [10]. Despite the different estimation equations, a universal model still does not exist due to the complexity of the process.

Continuing in this context, the authors in Refs. [11–14] highlighted the influence of land use on sediment concentration and load in the water body. The study by Zhang et al. [15] obtained results showing that higher pollution intensity and lower vegetation cover lead to higher sediment loads in the water body, thus becoming an aggravating factor concerning the previously mentioned problems.

The study by Garrido et al. [16] conducted 15 hydro-sediment metric measurement campaigns to determine the rating curve that represents the suspended sediment loads in the Piranhas River, located in the state of Paraíba, Brazil. This type of study assists in the planning and development of dam projects for multiple purposes.

Sediment transport in a natural river is a transient process due to changes in flow at any given moment caused by random events. Various models have been developed to estimate the amount of sediment transported, aiming to gain a deeper understanding of sediment transport dynamics and the events that influence this process [17].

Water and sediment discharges affect the environmental, economic and social aspects inherent to the population that depends on the natural resources of the watershed. Sediments and their impacts currently represent a major challenge for water resource management.

The Jordão stream, located in the Triângulo Mineiro region of the State of Minas Gerais, Brazil, is characterized by strong rapids, that leave its bed rocky and sections where the flow allows sediment deposition on its bed. This stream features characteristics such as waterfalls, which enable the construction of dams for electricity generation. Hence, it becomes an important site for sediment transport studies in the Brazilian Cerrado region.

2. Material and method

2.1 Characterization of the study area

The work was carried out in the river basin of the Jordão River located in the triangle of the Triângulo Mineiro and Alto Paranaíba in the Unit of Planning and Management of the Water Resources – high Paranaíba River 1 (UPGRH-PN1), between the geographic coordinates 18°25′41″ and 18°50′51″ south latitude and 47°49′47″ and 48°12′51"west longitude. The area of the basin is approximately 926 km^2 and the extension of the main waterway is approximately 90 km from the source to the confluence with the Paranaíba River.

The collection points were defined based on planning and study of the course of the river and the watershed. Initially the area was characterized by obtaining maps elaborated in the QGIS tool. The version used in this study is QGIS 2.18. QGIS is a user-friendly Open-Source Geographic Information System (GIS) licensed under the GNU General Public License [18]. Five collection points of water and sediment samples were established.

Figure 1 illustrates the location of the Jordão River basin, which is the object of study of this project.

Five fluviosedimentometric stations were installed along the course of the river. Among the several tributaries of the River Jordão, the Bom Jardim stream stands out for its direct influence on the river; the stream mouth is located between stations Pissarão 2 and Jordão, where it is possible to validate the contribution of this stream in relation to the liquid and solid discharges deposited. Collections were performed at points Pissarão 1 and 2, Bom Jardim, Jordão and Jordão Foz (station at the mouth of the Jordão River), where the length of the stretch between points Pissarão 1 and Jordão Foz is approximately 59 km (**Figure 1**). Point Jordão Foz is close to the mouth of the river, at the confluence of the Paranaíba River. The region under study is characterized by the presence of large agricultural areas, a significant amount of pasture areas, as well as vegetation and urbanized patches.

2.2 Hydrological and sedimentological monitoring

A total of eight campaigns were carried out with measurement and collection field activities performed in both dry and rainy seasons from September 2018 to May 2019. Between these periods, there was no field activity performed in the month of October of 2018. Flow measurements and water samples for the analysis of suspended sediments were performed from month to month. For bed sediments, two collections and a sieving analysis were carried out throughout the research performed at the five stations. The stations with their respective photos of the sites are represented in **Figure 2**. Due to access difficulties at the Jordão Foz station, the decision was made to end sampling at that location in January 2019.

The Acoustic Doppler Current Profiler (ADCP) or fluviometric windlass equipment was used to determine the cross-section, velocity profile, point velocity and net

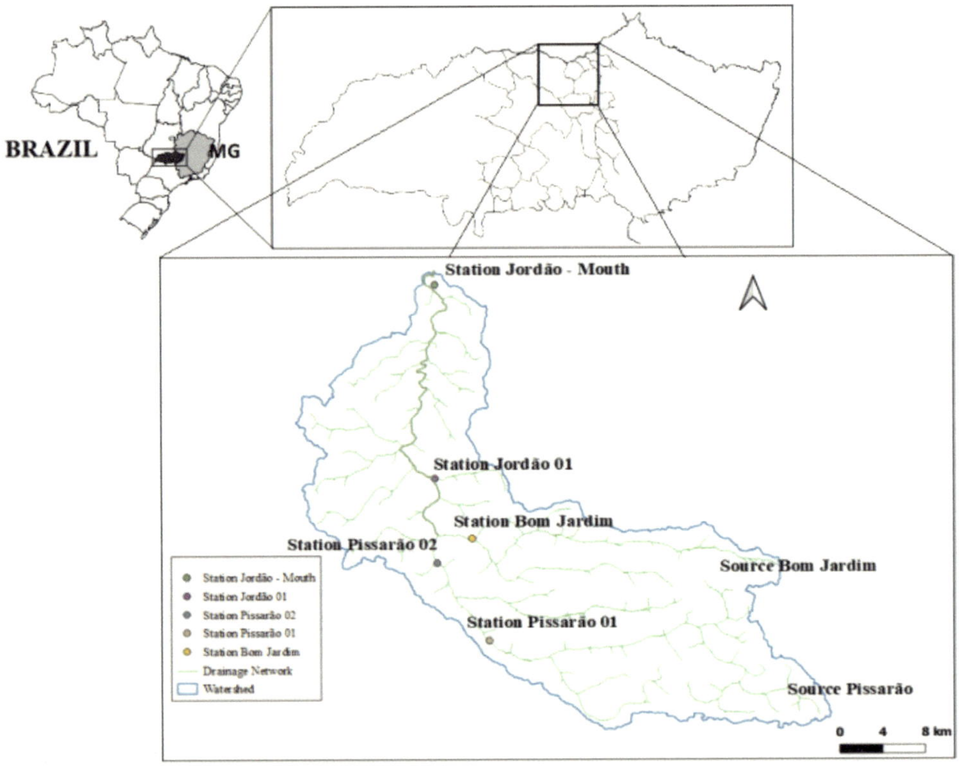

Figure 1.
Representative map of the location of the Jordão River basin. Source: The author [19].

flow rate of the watercourse. The Doppler effect is the fundamental principle used by the ADCP, which is provided with transducers (emitters and receivers); these are responsible for the emission of sound waves of fixed frequency, which, when encountering particles suspended in the water, these waves are reflected and detected by the transducers the moment the sound comes into contact with the particles. In the present research, the flow rates at each section were obtained through fieldwork.

In the field, a rope was tied to the ADCP that was connected to a notebook containing the River Surveyor software provided by the manufacturer. Then, with the aid of the rope, at least three crossings were carried out at each sampled section of the river. Regarding the profile from the bottom meter, the profiles and speed directions were recorded, and the flow calculation was performed by the software. The fluviometric windlass was used to measure flow rates on the sections when the water level was below 30 cm, and the ADCP was unable to record the flow rate. This device has a dynamic propeller that rotates under the action of river water flow, and the number of revolutions of this propeller is registered on an electronic revolution counter. In the procedure performed in the field, a graduated rope was used to perform meter-by-meter readings with the verticals fixed (forming vertical subareas) along the actual cross-section of the watercourse. The distance of 1 meter between the verticals was established according to the width of each section, as recommended by Santos et al. [20].

The collection of water for analysis of suspended sediment concentration was performed by the indirect method of sampling by equal width increase – IIL [10].

DOI: http://dx.doi.org/10.5772/intechopen.1011122

(a) Station Pissarão 01: 18°45'45,58"S 48° 2'53,22"W

(b) Station Pissarão 02: 18°41'33,02"S and 48°05'39,7"W

(c) Station Bom Jardim: 18°40'13.38"S and 48° 3'47.59"W

(d) Station Rio Jordão:18°35'20"S and 48°7'41" W

(e) Foz Jordão Station: 18°26"28.83"S and 48°05'43,99"W

Figure 2.
Photographs of the collection stations: (a) and (b) Stations on the Pissarão stream (a tributary of the Jordão stream); (c) stations on the Bom Jardim stream (a tributary of the Jordão stream); (d) and (e) points along the Jordão stream. Source: The author [19].

The procedure was performed with the aid of the North American series integrating sampler (U.S-DH 48). During sampling, the water and air temperature were measured using thermometers.

For sampling, each cross-section of the river was divided into segments of equal widths (i), using a graded rope with the selection of 10–20 verticals. Samples were collected at the same vertical velocity, obtaining a series of sub-samples with different volumes. In all collections, the same sampler was used with the diameter of the 1/8"collection nozzle.

The analyses of concentration and granulometry of suspended sediments were performed at the Laboratory of Sanitation (LABSAN) of the Faculty of Civil Engineering, as described by Santos et al. [21] in the ANEEL guide for sedimentometric practices. Upon arriving from the field, the samples were weighed to avoid errors in the calculations due to the possibility of water loss by evaporation. The sediment concentration analyses were performed by filtration methods in dry seasons with lower concentrations and evaporation in rainy seasons with higher concentrations.

All analyses were performed in the shortest time possible to avoid changes in the characteristics of the samples.

The suspended sediment concentration Css (mg/L) was obtained in Eq. (1), establishing the relationship between the mass of the sediment mss (mg) (dry sediment) and the volume of water with the collected sediment V (L):

$$Css = \frac{mss}{V} \tag{1}$$

With the determination of suspended sediments and liquid discharge measurements, the suspended solid discharge of each studied section was calculated. The solid discharge represents the amount of material transported in mass per unit of time according to Eq. (2):

$$Qss = 0{,}0864 \cdot Q \cdot Css \tag{2}$$

In which: Qss: suspended solid discharge (T/day); Q: liquid discharge (m^3/s); Css: Suspended sediment concentration (mg/L).

The particle size analysis on the bed sediment samples was carried out following the Brazilian Association of Technical Standards – ABNT [22] to identify the particle size or grain size and its distribution as a percentage of each grain diameter. The collected material was taken to the Geotechnical Laboratory and prepared according to Associação Brasileira de Normas Técnicas [23]. In order to determine the grain size distribution of the coarse sediment fraction (particles larger than 2.0 mm in diameter) and fines (particles smaller than 2.0 mm and greater than 0.075 mm), the sieving test was performed.

The apparatus and materials used in the test were: mechanical stirrer with sieves, # 50, # 38, # 25, # 19, # 9.5, # 4.8, # 2.0, # 1, 2, # 0.6, # 0.42, # 0.25, # 0.15 and # 0.075 mm, brush with metal bristles, scale, capsules, mortar and greenhouse.

Samples from each collection point were air dried to near hygroscopic moisture. The dried material was de-corroded and homogenized. Through the quenching treatment, the amount of material was reduced until a representative sample was obtained for performing the assay. For each sample, triplicate tests were performed with about 2.0 kg of material.

The Halley-Smith equipment was also used to estimate the direct sediment discharge at the upstream stations of the Jordão stream, namely Pissarão 1, Pissarão 2, and the Bom Jardim stream, with the purpose of comparing the results with the five indirect methods tested.

The equipment was fixed in three different positions along each cross-section, placed at the margins and in the central region. At each sampling point, it remained for 30 minutes, totaling 90 minutes of sampling per cross-section. During this time, sediment mass was collected at each of the study sections, and solid discharge was calculated.

2.3 Hydraulic and geometric characterization of the Jordão River

With respect to estimating sediment discharge in the Jordão River, it was necessary to obtain the hydraulic and geometric parameters of the analyzed sections. The flow

(Q), mean velocity (U), mean depth (d), width (L) and the bathymetric top profile of the sections were all previously determined in the field. The slope of the water line, roughness coefficient, perimeter, area and hydraulic radius were determined from the River Surveyor program and by calculations with established equations from the literature.

The confirmation of the water line slope (S) was performed by using the Manning Eq. (3). Manning's formula is the most widely used for determining the slope in rivers and canals due to proven results on its validity and the large amount of data derived from its overall application.

The roughness coefficient (n) was obtained from the mean of the Strickler formula changed to D_{50} and D_{90} (Eqs. (4) and (5)). To determine the wetted perimeter (Pm), the sections were analyzed from the profiles obtained in the River Surveyor program. The wet area (Am) was obtained by the continuity Eq. (6). The hydraulic radius was determined by the ratio of the wet area to the wetted perimeter:

$$S = \left(\frac{Q \cdot n}{A \cdot Rh^{2/3}} \right)^2 \tag{3}$$

$$n = \frac{D_{50}^{1/6}}{21,1} \tag{4}$$

$$n = \frac{D_{90}^{1/6}}{26} \tag{5}$$

$$Q = U \cdot A \tag{6}$$

2.4 HEC-RAS program

The HEC-RAS – River Analysis System program, currently in version 6.5, from March 2024, was developed by the Hydrologic Engineering Center for the U. S. Army Corps of Engineers.

In this study, HEC-RAS was used to simulate the transport of solid particles in two main tributaries, the Pissarão stream and the Bom Jardim stream, and after the confluence of the two in the Jordão stream to its mouth. Therefore, it was possible to analyze over the studied time, which was from September 2018 to May 2019, the places where erosion and sediment deposits occur.

During the simulations, the HEC-RAS program, in its integrated system, allowed for the manipulation and visualization of the input data, the flow simulation, in a so-called almost permanent regime (sedimentary transport on a moving bottom), along with analysis of the results on tables and graphs, in addition to the reporting. The calculation and graphic presentation were conducted in a so-called almost permanent regime, in a network of channels with a natural section.

For the simulation in the HEC-RAS model, it was necessary to introduce basic information of different types from the field surveys obtained during the collections carried out, which include ((a) Geometric data obtained during flow measurements using ADCP equipment, including the spacing between cross sections; (b) hydraulic parameters, such as the Manning roughness coefficient; and (c) the selected method for sediment transport analysis at the study cross sections. These data, along the simulated stretch in five sections, were added manually. The other sections were obtained using the ArcGIS tool. The sections were added every 5000 meters along the sequence, and interpolated sections were created between those already existing, using the existing interpolator function in HEC-RAS. After interpolation, the maximum distance between the two sections went from 5000 to 500 meters. The Manning coefficient was determined for the stations from the granulometry data by using Eqs. (3) and (4), as indicated in the results.

The computational procedure is based on the solution of the one-dimensional energy conservation equation. Pressure losses are taken into account through friction (Manning's equation) and through contraction/expansion (product of the contraction/expansion coefficient due to the variation in kinetic height). The equation of conservation of momentum is used in situations where the backwater curve varies very quickly. These situations occur when there is a change in the flow regime (e.g., in the hydraulic ledge), when crossing bridges and at the confluences of river stretches.

After creating the geometry, the flow data obtained in the collections were inserted as boundary conditions for the quasi-permanent flow. To enter the data, the nascent stations were firstly considered, respectively, in the Bom Jardim stream and in the Pissarão stream as the upstream limit. The downstream boundary condition, next to the mouth of the Jordão stream, was considered as the slope of the water line in the last stretch. In the flow series editor, flow data, flow duration and computational increment were entered. The flow rates were those obtained in each campaign; the flow duration consists of the difference in hours from one collection to another and the computational increment was considered as 24 hours.

Before inserting the data of bottom and suspended sediments and the watershed, tests were carried out with simulations of the hydraulic model, since the sediment results are very sensitive to hydraulic parameters.

To enter sediment data into the program, the granulometric distribution of the bed of the five stations was first inserted from the analysis and granulometric curves created in the campaigns. These analyses were previously carried out in the field, along with the predominant soil type in the basin and the concentration of suspended sediments according to the flow measured in the field. The granulometric distribution in each section of the program was considered as being that of the closest station to where bed sediments were collected and analyzed, and the same was performed for suspended sediments. In all sections, the definition was made that the maximum change in the vertical dynamics of the bed would be 0.5 m.

The simulations in the program were defined between the months of September and May, using the Ackers and White methods for the sediment transport (transport function) for the classification method in order to calculate the thickness of the active layer in the "Sorting Method," where there are two options defined: Exner 5 and Active Layer. The first limits bed erosion, and the second is indicated for gravel layers, being used only when the Wilcock equation is chosen; finally, the Toffaleti method was defined for sedimentation speed. The results obtained were divided into two types of graphs, the first being the record of the total change in the bed mass accumulated over time and in space (a). The period was defined as the sampled period from September 1, 2018, to May 31, 2019, and the space between the beginning of the cross-section until just before the upstream section of the first section. The second graph shows the transport with drag or deposition of sediments in the cross-section over time (b). The decision was made to generate sections next to the source and mouth of each studied tributary, as well as on the Jordan River, which is the junction of the Pissarão and Bom Jardim streams.

3. Results and discussion

In this section, all the results obtained are carefully analyzed for sediment transport in this watershed and accordingly presented. Initially, the data collected during each fieldwork campaign, including the hydraulic and geometric data of the sections, are presented in **Tables 1–5**.

Analyzing all the Hydraulic and Geometric Data Tables, the wide variation in the characteristics of each section studied is evident. The values Q of flow in the river increased significantly in the season of sharp precipitations, which was already expected.

In the dry period, the flow rates determined at the different stations were in the range of 0.392 to 4.08 m^3/s. However, in the rainy months, the flow rates were of 0.908 to 11.096 m^3/s. In addition to the flow, the data of width, depth, hydraulic radius, slope and velocity were used to calculate sediment discharge.

Regarding the mean depth, the lowest values of d were identified in section Bom Jardim, with an average value of 0.175 m. The highest depth values of d were obtained in section Jordão Foz, with an average of 0.829 m. In the rainy season, the depths were significantly higher, especially in the Jordão section, where there was a variation of up to 0.315 m.

The highest value U of mean flow velocity was obtained in section Bom Jardim, with a value of 2.274 m/s. Sections Pissarão 02 and Jordão presented the lowest flow velocities with averages of 0.415 and 0.38 m/s, respectively. In the rainy season, the velocities increased across all sections; however, section Bom Jardim presented the most significant difference, with an average value of 905% higher in the rainy season when compared to the dry season.

Date	Period	Station Pissarão 1							
		L (m)	d (m)	U (m/s)	Q (m^3/s)	A (m^2)	Pm (m)	Rh (m)	S (m/m)
04/09/2018	Dry	5.63	0.196	0.241	0.549	2.277	6.035	0.377	0.00006
20/11/2018		7.65	0.417	1.171	3.293	2.813	7.836	0.359	0.00155
11/12/2018		7.10	0.392	0.803	2.824	3.516	7.273	0.483	0.00049
23/01/2019	Rainy	6.47	0.403	0.555	1.638	2.951	6.755	0.437	0.00027
27/02/2019		6.57	0.397	0.645	2.052	3.183	6.841	0.465	0.00033
20/03/2019		6.14	0.383	0.675	2.071	3.066	6.702	0.457	0.00037
20/04/2019		6.44	0.395	0.504	1.525	2.830	6.640	0.412	0.00027
22/05/2019	Dry	6.04	0.297	0.375	1.047	2.562	6.410	0.396	0.00017
Average		6.57	0.360	0.621	1.875	2.899	6.812	0.423	0.00044

L: width; d: mean depth; U: average water velocity; Q flow; Am: Wet area; Pm: Wetted perimeter; Rh: Hydraulic radius; S: slope.

Table 1.
Hydraulic and geometric characteristics of station Pissarão 1.

Date	Period	Station Bom Jardim							
		L (m)	d (m)	U (m/s)	Q (m^3/s)	A (m^2)	Pm (m)	Rh (m)	S (m/m)
04/09/2018	Dry	5.52	0.292	0.251	0.511	2.034	6.140	0.331	0.00008
20/11/2018		6.71	0.093	1.511	1.989	1.316	7.614	0.173	0.00685
11/12/2018		6.68	0.083	2.274	1.844	1.211	7.001	0.116	0.02641
23/01/2019	Rainy	5.61	—	—	1.014	—	—	—	—
27/02/2019		6.86	0.114	1.604	1.675	1.044	6.870	0.152	0.00917
20/03/2019		5.49	0.123	2.321	2.040	0.879	5.803	0.151	0.01936
20/04/2019		5.46	0.236	0.831	0.908	1.605	5.901	0.272	0.00052
22/05/2019	Dry	5.45	0.281	0.245	0.463	1.890	5.940	0.318	0.00008
Average		6.02	0.175	1.291	1.347	1.413	6.467	0.218	0.00892

L: width; d: mean depth; U: average water velocity; Q flow; Am: Wet area; Pm: Wetted perimeter; Rh: Hydraulic radius; S: slope.

Table 2.
Hydraulic and geometric characteristics of station Bom Jardim.

Date	Period	Station Pissarão 2							
		L (m)	d (m)	U (m/s)	Q (m^3/s)	A (m^2)	Pm (m)	Rh (m)	S (m/m)
04/09/2018	Dry	14.5	0.275	0.180	0.392	2.177	15.022	0.145	0.00012
20/11/2018		15.5	0.583	0.521	4.786	9.193	16.295	0.564	0.00017
11/12/2018		15.6	0.570	0.523	4.598	8.792	16.287	0.540	0.00018
23/01/2019	Rainy	15.2	0.445	0.429	2.882	6.697	16.032	0.418	0.00017
27/02/2019		15.3	0.501	0.481	3.910	8.091	16.392	0.500	0.00017
20/03/2019		15.2	0.467	0.459	3.463	7.528	16.318	0.463	0.00017
20/04/2019		—	—	—	—	—	—	—	—
22/05/2019	Dry	15.0	0.356	0.310	1.585	4.142	15.478	0.266	0.00025
Average		15.2	0.457	0.415	3.090	6.660	15.975	0.414	0.00018

L: width; d: mean depth; U: average water velocity; Q flow; Am: Wet area; Pm: Wetted perimeter; Rh: Hydraulic radius; S: slope.

Table 3.
Hydraulic and geometric characteristics of station Pissarão 2.

Regarding slope, station Jordão had the lowest values of S with an average of 0.00003 m/m. However, the highest values of S remained in section Bom Jardim. This parameter is very important as it directly influences sediment transport along the river.

The characterization of the granulometry of bed sediments was essential for the understanding of the granulometric distribution of the particles in the river.

Figure 3 shows the grain size curves of the stations.

Analyzing particle size curves over the sections, it becomes evident that the particles on the stream bed are basically composed of sand and gravel.

The granulometric curves indicate that between the dry and rainy periods, there was an increase in the diameter of the sediments present on the stream bed, especially in the Pissarão stream and at the mouth of the Jordão stream. In the Bom Jardim stream, this assertion is not so clear-cut, as the results from the bed sediment collection for the

Date	Period	Station Furnas							
		L (m)	d (m)	U (m/s)	Q (m^3/s)	A (m^2)	Pm (m)	Rh (m)	S (m/m)
04/09/2018	Dry	13.5	0.590	0.250	1.834	7.336	14.892	0.493	0.00005
20/11/2018		15.0	0.720	0.453	9.256	20.437	16.445	1.243	0.00004
11/12/2018		15.0	0.729	0.460	9.142	19.870	16.500	1.204	0.00005
23/01/2019	Rainy	14.0	0.863	0.363	6.575	18.040	15.727	1.147	0.00003
27/02/2019		15.0	0.905	0.402	8.315	20.717	16.810	1.232	0.00004
20/03/2019		14.7	0.880	0.398	7.569	18.957	16.447	1.152	0.00004
20/04/2019		14.5	0.856	0.353	5.871	14.700	15.773	0.898	0.00004
22/05/2019	Dry	14.0	0.804	0.325	4.080	12.406	15.592	0.793	0.00004
Average		14.5	0.790	0.380	6.580	16.560	16.020	1.020	0.00004

L: width; d: mean depth; U: average water velocity; Q flow; Am: Wet area; Pm: Wetted perimeter; Rh: Hydraulic radius; S: slope.

Table 4.
Hydraulic and geometric characteristics of station Jordão.

Date	Period	Station Foz							
		L (m)	d (m)	U (m/s)	Q (m^3/s)	A (m^2)	Pm (m)	Rh (m)	S (m/m)
04/09/2018	Dry	16.5	0.559	0.429	2.543	5.928	17.386	0.341	0.00005
20/11/2018		18.5	0.984	0.572	10.030	17.535	19.993	0.877	0.00004
11/12/2018	Rainy	18.5	0.993	0.601	11.096	18.463	20.091	0.919	0.00005
23/01/2019		16.7	0.780	0.489	7.625	15.578	17.842	0.873	0.00003
Average		17.55	0.829	0.523	7.823	14.376	18.828	0.753	0.00004

L: width; d: mean depth; U: average water velocity; Q flow; Am: Wet area; Pm: Wetted perimeter; Rh: Hydraulic radius; S: slope.

Table 5.
Hydraulic and geometric characteristics of station Jordão Foz.

month of September 2018 are not available. This may occur due to the transport of the fine particles from the high course to the low course of the river. Research by Brigante and Espíndola [24] also confirmed this fact in their research work.

At the Jordão station, after the confluence of the Pissarão and Bom Jardim streams, with the onset of the rainy season, an increase in fine sediments is noted in the results in **Figure 3** and to a lesser extent, coarser sediments were observed, which possibly originated upstream from the adjoining waterways. This cross-section is characterized as a sediment deposition region, which will be discussed further in this chapter.

At the mouth of the Jordão stream, the quantity of sediments between 0.3 and 20 mm was reduced, indicating sediment transport downstream from this station. Further downstream, an increase in larger-diameter sediments is observed. One hypothesis to explain this occurrence is the transport of finer sediments, which reveals the presence of coarser sediments that were possibly located beneath finer layers and have become exposed and subject to collection with the Petersen sampler.

According to Brigante and Espíndola [24], the turbulence provoked by the flow of water in the upper reaches of the rivers has a velocity capable of suspending the finer

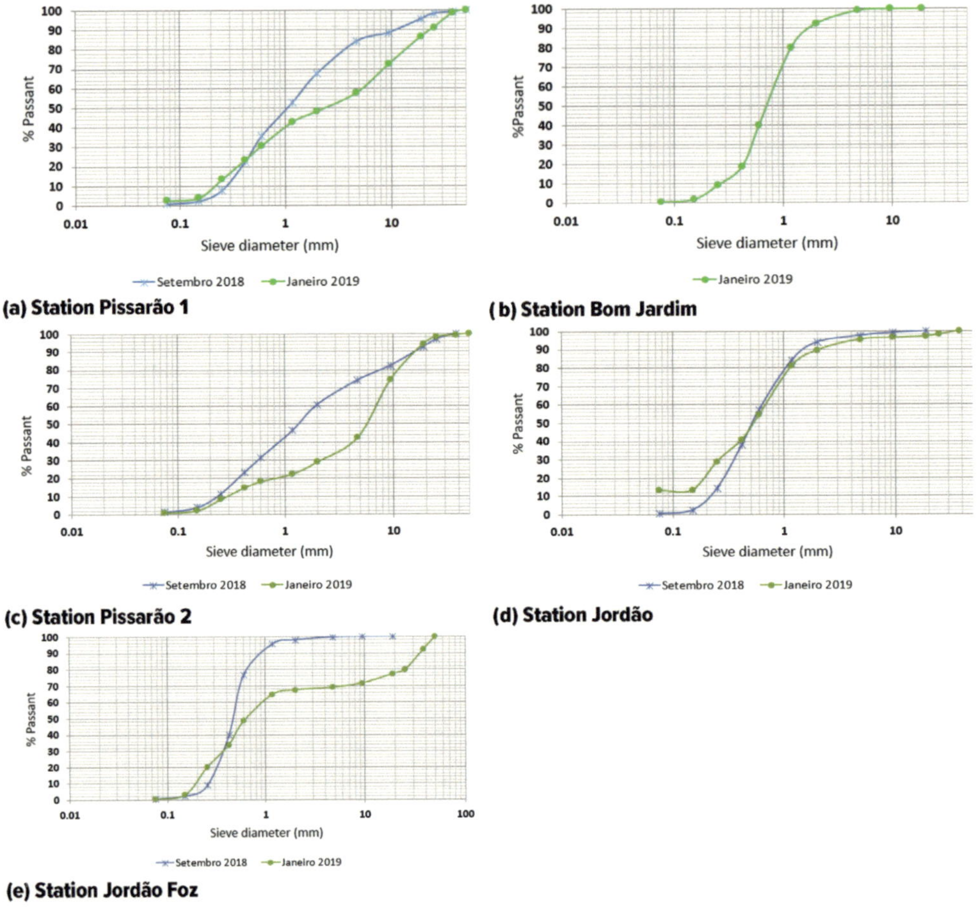

Figure 3.
Granulometric curves of the bed material: (a) Station Pissarão 1; (b) Station Bom Jardim; (c) Station Pissarão 2; (d) Station Jordão and (e) Station Jordão Foz.

fractions of the sediment, transporting these along the course of the river toward the mouth. Therefore, the tendency is that the sites closest to the river mouth have background sediments with smaller particles than the upstream sites. This fact is evidenced in the Jordão River, when station Pissarão 1, upstream, is compared to station Foz, near the mouth, it is clear that the bottom material of section Foz is composed of finer particles, especially in the period preceding the rains, in this case, the month of September.

After the onset of the rainy season at all stations, in general, with the transport of finer sediments, the riverbed was left with coarser sediments that were not carried downstream.

Based on the information contained in **Tables 1-5** and **Figure 3**, it was possible to estimate the sediment discharge using five indirect estimation methods found in Beling et al. [8] (Colby [25], Ackers and White [26] and Engelund and Hansen [27]). In presenting the results, graphs were constructed with the results of the indirect methods and the liquid discharges measured with the ADCP at each station separately, as shown in **Figure 4**, where on the x-axis, number 1 indicates the month of September. Number 2 indicates October and so on.

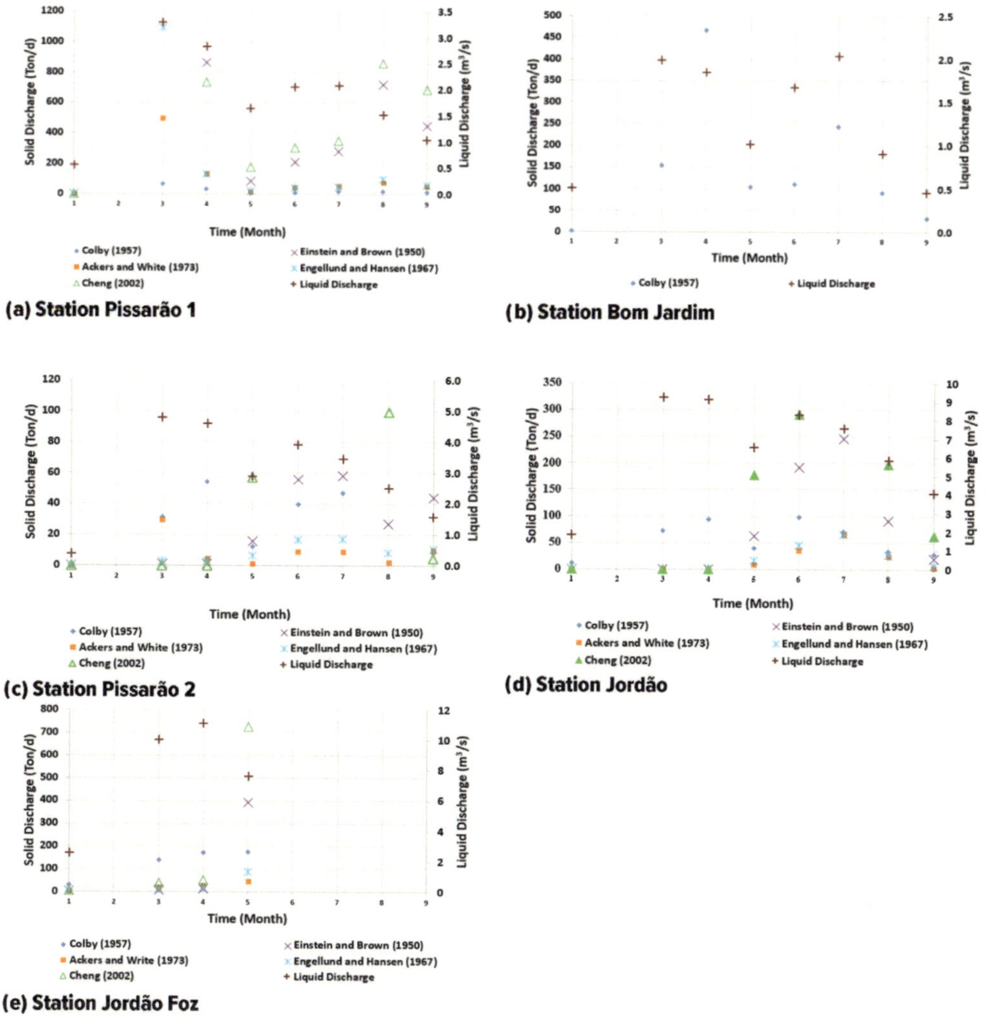

Figure 4.
Indirect Methods Results: (a) Station Pissarão 1; (b) Station Bom Jardim; (c) Station Pissarão 2; (d) Station Furnas and (e) Station Furnas Foz.

When considering the results obtained with the Halley-Smith equipment, the methods that best match the estimated sediment discharge are those presenting lower values when compared to all the methods. Consequently, the methods of Colby [25], Ackers and White [26] and Engelund and Hansen [27], respectively, provide the most coherent results. However, in November, the latter two methods showed discrepancies precisely with the onset of the rainy season and the consequent increase in liquid discharge, which returned to normal in the following months.

In the study by Brigante and Espíndola [24], analysis of variance (ANOVA) was applied to these results, which indicated that the means do not differ from each other at the Pissarão 1, Pissarão 02 and Jordão stations, suggesting a reliable relationship between the methods at these stations. The same was noted at the Jordão Foz station, located near the mouth of the Jordão stream. However, the F-result was shown to be very close to the critical value, which may lead to unreliable results.

Finally, at the station located on the Bom Jardim stream, there was no significance among the five methods used. As will be discussed later, the Colby [25] method produced more coherent results at this data collection station.

It is still not possible to determine which conditions caused the methods to present differences in the results between the stations, especially at the Pissarão mouth and Bom Jardim sampling stations. One hypothesis could be the hydraulic characteristics or the seasonality of sediment discharge, which could indicate the need to change the cross-section in at least two sampling stations.

By analyzing the graphs for sediment discharge and liquid discharge over time, one notes that the highest intensity peaks occur mainly in the period with higher rainfall, as is the case in December. In the dry months, such as September, there is a marked decrease in the flow of suspended solid discharge.

Noteworthy from the hydraulic and geometric parameters of the sections (**Tables 1–5**) is that conclusions can be reached on some important correlations with the data for flow, concentration and suspended solid discharge. Station Jordão and Foz, for example, had the highest values for sediment concentration, especially in the rainy periods. This is a consequence of the increase in the turbulent flow regime, which facilitates the resuspension of particles that were deposited in large quantities on the riverbed. This region presents a high accumulation of bed sediments due mainly to the low slope and, consequently, the low flow velocities. Even in this season, it is interesting to note that the average hydraulic radius value is higher than the other sections, showing the tendency of this region to present a greater wet area when compared to other stations, thus configuring a flow that interferes less with the background particles and therefore possesses less influence concerning the transport of these particles.

In stations Pissarão 1 and Bom Jardim, steep slopes and flow velocities are observed, with lower mean hydraulic radius values. At these stations, the finer particles, in particular, are greatly influenced by the flow forces causing the transport of these particles to lower regions, promoting an increase in the bottom sediment load, especially at stations Jordão and Foz.

Suspended sediment discharge and discharge lines are similar across all seasons and the periods of these variables coincide in most months. Regarding changes and leaks, these do not show the same characteristics since there is significant variation between lines and the numbers coincide for only a few months. In section station Jordão, for example, a concentration line shows high-intensity peaks in January and April 2019, while the peak leakage is displayed only as it occurs in the months preceding this period.

According to Glysson [28] and Carvalho [11], the time between peak sediment concentration and peak leakage can dramatically affect sediment curves due to the large number of dispersion points. Therefore, as shown in the result graphs, a variation in the monthly period of the directly affected peaks or in the key settings, as the peaks in the monthly period seem to be the case of sediment leakage and discharge produced by coefficient methods larger than those which occur for leakage and concentration that showed peaks at different times, thus evidencing the statement made by the author.

The Halley-Smith equipment was also used to estimate the direct sediment discharge at the upstream stations of the Jordão stream, namely Pissarão 1, Pissarão 2 and the Bom Jardim stream, with the purpose of comparing the results with the five indirect methods tested.

The rainy season was chosen for the study as the assumption was made that a greater quantity of transported sediments would occur. However, even during periods of higher flows, the measured sediment quantity was almost negligible. The range of sediment diameters collected by the equipment was between 0.08 and 2 mm. According to Brazilian standards, these are classified as sand, comprising fine, medium and coarse fractions.

The granulometric results of the collected material are presented in **Figure 5**, which shows the granulometric curves discussed later. It is noteworthy that these curves differ from those presented in **Figure 3**, as they represent sediments collected while in motion along the creek beds, unlike the previously collected sediments that were fixed to the bed. The estimation was also made possible for the sediment quantities and the respective solid discharges, which are presented in **Table 6**.

However, the indirect methods showed higher results than those obtained with the Halley-Smith equipment. Among the three tested sections, the methods of Ackers and White, Engelund and Hansen and Colby, in that order, can be considered the most representative of sediment transport in this river. The first two methods are based on bedload sediment transport, while the Colby method is designed for suspended sediment.

At Pissarão 1 station, during November and December, the transported sediments remained practically the same, while in January, there was a slight reduction in particle size, indicating changes in the riverbed due to variations in flow velocity and the downstream transport of these particles.

At Bom Jardim station, from November to December, there was a reduction in the quantity of finer sediments and an increase in coarser sediments. However, in January, the same trend was noted at Pissarão 1 station, with a reduction in particle sizes but with more pronounced values.

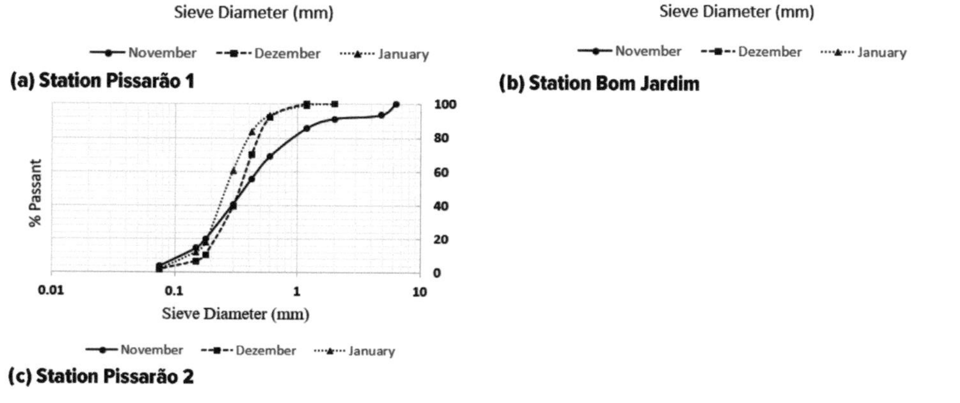

Figure 5.
Granulometric curves of the bed material collected with the equipment. Halley-Smith: (a) Station Pissarão 1; (b) Station Bom Jardim and (c) Station Pissarão 2.

Season	Discharge solid (kg/day)		
	November	December	January
Pissarão 01	2.65	1.33	0.13
Bom Jardim	3.47	5.34	0.77
Pissarão 02	53.94	30.07	19.17

Table 6.
Sediment Mass collection using the Halley-Smith equipment at stations.

Finally, at station Pissarão 2, located downstream of Pissarão 1 and upstream of the confluence with Bom Jardim creek, observations showed that, on average, the quantity of fine sediments (0.06–0.3 mm) remained constant over the 3 months of study, while the coarser sediments (0.3–2 mm) decreased from November to December and remained practically constant in January.

The solid discharge measured during each field survey with the Halley-Smith sampler is presented in **Table 6**.

Table 6 shows that these values are very small. Therefore, to understand the dynamics of sediment collection in motion along the streambeds, the graphs in **Figure 6** were created, illustrating the trends observed in sediment transport during the period when samples were collected using the Halley-Smith sampler.

Noteworthy here is that at the three stations, the results presented and compared to the methods of Colby, Ackers and White, Engellund and Hansen and Cheng were valid in terms of growth and decrease relationships concerning solid discharge.

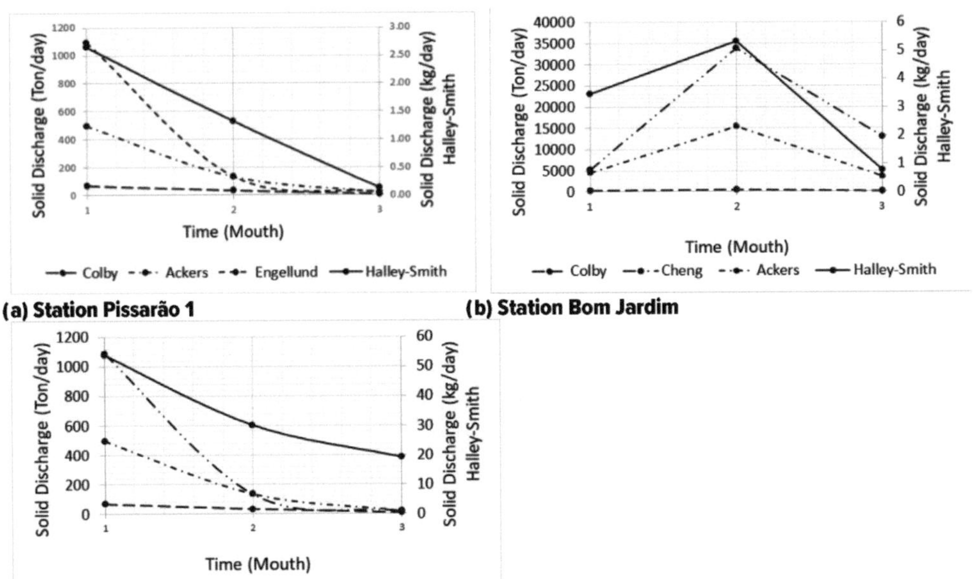

Figure 6.
Comparison between the results obtained by indirect methods and the Halley-Smith equipment. (a) Station Pissarão 1; (b) Station Bom Jardim; (c) Station Pissarão 2.

Regarding stations Pissarão 1 and 2, there was a decrease between November and December, while at station Bom Jardim, there was an increase between November and December and a decrease in January compared to December. This indicates at least a certain relationship, which should be further verified through additional fieldwork.

In order to understand the movement of sediments in the cross-sections along the entire stretch of the Pissarão and Bom Jardim streams, as well as the Jordão River formed by the confluence of the first two, sediment transport simulations were carried out using the HEC-RAS program, following the same procedures described in Pereira et al. [19].

The simulations using HEC-RAS resulted in sediment load estimates for the three studied reaches, as shown in **Table 7a-c**, which present regions with predominant erosion and others with sediment deposition. The cross-sections in these tables are arranged from the source to the mouth, allowing for visualization of sediment transport starting from the source of each tributary and continuing along the Jordão River after the confluence of these two streams.

During the period under analysis, it became evident that most of the sections experienced bed erosion in the Bom Jardim stream, regardless of whether it was the dry or rainy season, with sediment deposition occurring only at Section 32326.

Cross-section	**Simulation time**	**Discharge**
(m)	**(days)**	**(Ton/day)**
(a) Bom Jardim		
104,584	01/09/2018 to 31/05/2019	0
61,970	01/09/2018 to 31/05/2019	−57.72
32,326	01/09/2018 to 31/03/2019	95.79
32,326	01/04/2019 to 31/05/2019	−41.97
11,886	01/09/2018 to 07/12/2018	−11.15
(b) Pissarão		
156,310	01/09/2018 to 31/05/2019	0.00
120,056	01/09/2018 to 31/05/2019	−44.24
98,406	01/09/2018 to 31/05/2019	−35.88
75,522	01/09/2018 to 31/05/2019	−10.24
54,680	01/09/2018 to 31/05/2019	40.95
35,530	01/09/2018 to 31/05/2019	−148.70
21,598	01/09/2018 to 31/05/2019	178.66
8202	01/09/2018 to 31/05/2019	−74.96
(c) Rio Jordão		
76,166	01/09/2018 to 31/05/2019	−47.18
50,314	01/09/2018 to 31/05/2019	164.86
24,612	01/09/2018 to 31/05/2019	−742.55
1432	01/09/2018 to 30/11/2018	2043.79

Table 7.
Results obtained from numerical simulations using the HEC-RAS software by cross-section for each stream (córrego) and river (rio).

Section 11886, the closest to the data collection point named Bom Jardim, showed erosion during the dry period, which intensified with the onset of the rainy season. In the overall sediment balance for this stream, erosion was predominant, with 6600.84 tons of sediment being removed during the simulation period and transported to the Jordão River.

In the Pissarão stream, near the headwaters and the Pissarão 01 data collection point, erosion occurred, ranging from 10.24 tons per day to 44.24 tons/day. At Section 54680, sediment deposition occurred, followed by significant erosion along the adjacent section, with high deposition values compared to upstream sections. At Section 8202, near the Pissarão 02 collection point, erosion of 74.96 tons/day was observed. Considering the overall sediment balance, the simulation indicates erosion in the stream totaling 25,680.59 tons, which will be transported downstream after the confluence.

The confluence of these two streams forms the Jordão River, and the simulations indicate that the amount of sediment arriving at this river from the two tributaries is 32,281.43 tons, which are initially deposited in Section 112449 in HEC-RAS. The first data collection section in this river is called Jordão 01 and is located near sections where visual observation shows that sediment deposition is predominant, especially after bends and areas of low slope in the Jordão River. At simulation Section 76166, erosion of the riverbed is observed, followed by deposition in the next section. In the last 30 km of the Jordão River, there is a strong current with the presence of meanders, which may explain the erosion of 742.55 tons/day, a value that is significantly different from the measurements taken in the upstream sections of this river. This location is upstream from the Jordão 02 data collection station, which is near the confluence with the Paranaíba River. At this station, the simulation showed that during the dry period, sediment deposition of 2043.79 tons/day occurred, and after the onset of the rainy season, these sediments were eroded at a rate of about 105.7 tons/

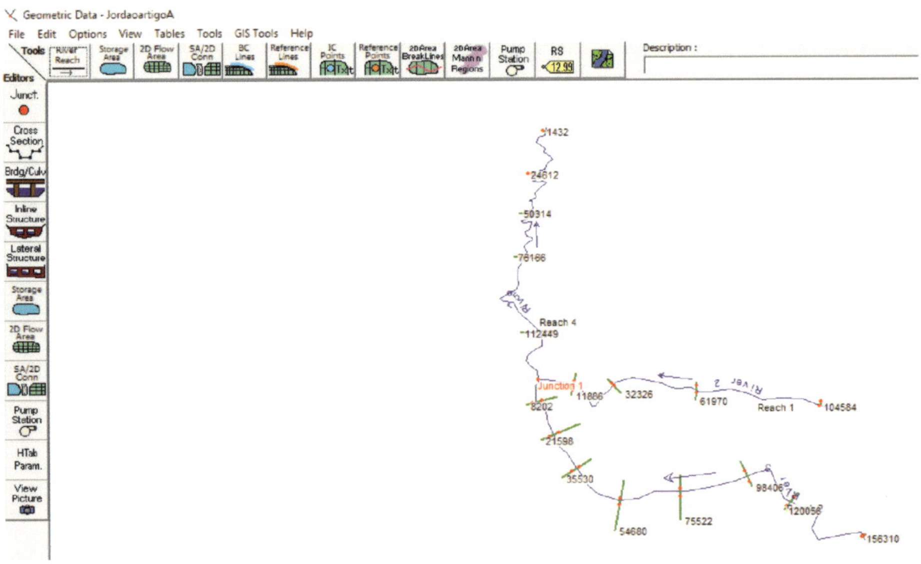

River 3 – Pissarão Stream; River 2 – Bom Jardim Stream; River 5 –Jordão river

Figure 7.
Image extracted from the HEC-RAS program showing the Pissarão tributary at the bottom, Bom Jardim above and after the confluence, the River Jordão.

day, much lower than the amount deposited. Since the liquid flow measurements were made with the ADCP equipment, it can be confirmed at the Jordão 02 collection site that the cross-sectional study area changed considerably between the dry and rainy periods, which helps corroborate these simulation-based measurements. Finally, it was found that the Jordão River contributed 31,506.60 tons of sediment to the Paranaíba River during the period analyzed in this simulation.

In general, the simulation herein aids our understanding of sediment transport along the studied watershed, including across the various cross-sections along the watercourses, where variations between erosion and sediment deposition occur. However, after sediment collections with the Halley-Smith equipment, it was observed that sediment transport occurs in a less pronounced manner than that observed in the simulation results. Therefore, there is a need for further studies and data collection to support and refine these numerical simulations.

Highlighted also was that the method used in the HEC-RAS results was that of Ackers and White [26], previously calculated and added to the graphs in **Figure 4**. However, in the simulations, a method for sedimentation velocity, namely Soulsby, was considered, which was not considered in the previous calculations (**Figure 7**).

4. Conclusions

The hydraulic and geometric characteristics obtained between September 2018 and May 2019 in the Jordan River basin showed variations between the fluviometric and sedimentometric stations and the sampling periods. The Pissarão 1 and Bom Jardim stations presented the highest flow velocities, while the Jordão station presented the lowest velocity values, and the two stations present on the Jordão stream presented the lowest average stream gradients.

The granulometric characterization of the stream bed sediments was essential for understanding the distribution of particles along the bed. The granulometric curves indicated that the material on the stream bed, along the studied sections, is basically composed of sand and gravel.

The granulometric curves constructed with the sediments collected using the Halley-Smith equipment indicated the diameter of the sediments transported along the watercourses with a predominance of sand, and at Pissarão 01 Station, there was a reduction in the diameters, indicating that finer diameters were transported downstream. In Bom Jardim, the thicker sediments were transported and at Pissarão 2 station the transport of finer sediments was noted with the beginning of the rainy season in the region.

Suspended sediment concentrations also showed variations between seasons and sampling periods. In the rainy season, sediment discharge was greater than in the dry season, highlighting the considerable influence of precipitation on the amount of sediment in the river. Between the stations, the highest average concentration was obtained along the Jordão section, and the lowest, in the Pissarão 2 section.

In addition to precipitation, factors such as gradient and vegetation cover influence the dynamics of sediments in the Jordão stream. The removal of vegetation cover directly affects the quantity and transport of particles in the river. The intensification of sediment discharge during periods of greater precipitation may be a consequence of the large amount of agricultural and pasture areas along the river basin. The steep gradient in the high regions of the studied section promotes the transport of sediments to the stations located along the Jordan River.

A preliminary understanding of the dynamics of sediments in the Jordan River basin, obtained through this research, is essential in its contribution to studies related to reservoir sedimentation, and construction of PCHs, as highlighted in Cobaner et al. [29], in addition to providing support in decision-making, which together help to improve the management of water resources in the basin. This dynamic included simulations carried out with the HEC-RAS program.

Finally, the need to continue research and work based on this hydrographic basin is highlighted, as there still exists obstacles when it comes to developing work related to sedimentology. In the current context, the use of numerical modeling, combined with reduced models, fieldwork and laboratory tests, is essential for more in-depth research and for a reliable understanding of this dynamic.

Acknowledgements

The author is grateful to Fundação de Amparo à Pesquisa do Estado de Minas Gerais (FAPEMIG), through grant number 3899/2016, for supporting this research and Faculdade de Engenharia Civil da Universidade Federal de Uberlândia.

Conflict of interest

The authors declare no conflict of interest.

Author details

Carlos Eugenio Pereira
Faculty of Civil Engineering, Federal University of Uberlândia, Uberlândia, Minas Gerais, Brazil

*Address all correspondence to: cepereira@ufu.br